T0185986

Programming and Interfacing with Arduino

Programming and Interfacing with Arduino

Dr. Yogesh Misra

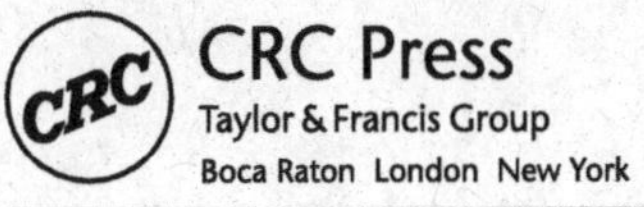

First edition published 2022
by CRC Press
6000 Broken Sound Parkway NW, Suite 300, Boca Raton, FL 33487-2742

and by CRC Press
2 Park Square, Milton Park, Abingdon, Oxon, OX14 4RN

CRC Press is an imprint of Taylor & Francis Group, LLC

ISBN: 978-1-032-05985-3 (hbk)
ISBN: 978-1-032-06316-4 (pbk)
ISBN: 978-1-003-20170-0 (ebk)

Typeset in Times
by codeMantra

Contents

Preface

This book provides a platform to the beginners to get started with the development of application by using Arduino UNO board. The objective of this book is to provide programming concepts of Arduino UNO board along with the working and interfacing of sensors, input/output devices, communication modules, and actuators with Arduino UNO board.

Arduino is an open-source hardware, which can be used to develop embedded systems with the help of open-source software. Arduino has gained huge popularity among the students and hobbyists for making a working model. The reasons behind the popularity of Arduino are its low cost, availability of software, and easy-to-interface possibility.

When someone is working with Arduino, he needs knowledge of three domains. First, he must understand the Arduino hardware board. Second, he must understand the Integrated Development Environment (IDE) required for the development of software, which actually guides the hardware to perform the desired task. Third, he must understand the working principle of various sensors, input and output devices, and actuators, which may be required to gather information from the surrounding for the processing by Arduino. The contents of this book are developed in keeping of the view of providing all information which is required for enhancing the expertise of all the domains required for the development of prototypes by using Arduino and associated peripherals. This book will be helpful in the development of employability skills in engineering undergraduate students.

After carefully understanding the exact requirements of the students and beginners, I am quite confident that easy-to-understand language of this book will make them efficient to learn Arduino. An outstanding and distinguished feature of this book is large number of programs with description and interfacing diagram associated with each program.

BOOK ORGANIZATION

This book starts with the explanation of Arduino UNO board and Integrated Development Environment (IDE). Various constructs required for the development of software are also covered. The working principle of various sensors is explained in depth; programming and interfacing examples with Arduino are taken up and finally some moderate-level projects.

Chapter 1 "Introduction to Arduino UNO Board" gives a detailed information regarding various components mounted on Arduino UNO board, IDE, and ATmega328 microcontroller.

Chapter 2 "Arduino Programming Constructs" gives an in-depth understanding of various constructs required for the programming of Arduino. Readers will also learn in this chapter about various operators, data types, and functions, which will be helpful to them when they start developing software for some specific applications.

Chapter 3 "I/O Devices, Actuators, and Sensors" deals with the construction and working of various sensors, input devices, output devices, and actuators. The readers will learn in this chapter the working principle and function of various pins of LED, seven-segment display, liquid crystal display (LCD), temperature sensor (LM35), humidity and temperature sensor (DHT11), light-dependent register, touch sensor, smoke detector (MQ2), rain detector (FC-07), ultrasonic sensor (HC-SR04), soil moisture sensor (YL-69), Bluetooth module (HC-05), GSM module (SIM 900A), switch, keypad matrix, potentiometer, analog-to-digital converter IC, motor driver board (L293D), and relay board.

Chapter 4 "Interfacing and Programming with Arduino" gives detailed information about how to interface input and output devices, viz., switch, keypad matrix, LED, seven-segment display, liquid crystal display (LCD) with Arduino UNO board, and the process of developing application programs for the interfaced circuit.

Chapter 5 "Arduino-Based Projects" covers interfacing and programming concepts with a large number of circuit diagrams of few projects based on Arduino.

All efforts have been made to keep this book free from errors. I sincerely feel that this book proves to be useful and helpful to the students for understanding the interfacing and programming of Arduino. Constructive criticism and suggestions from faculty members and dear students will be highly appreciated and duly acknowledged.

Dr. Yogesh Misra

Acknowledgments

I would like to express my deep sense of gratitude to Dr. Girish J, Director Education, GMR Varalakshmi Foundation; Dr. C.L.V.R.S.V Prasad, Principal, GMR Institute of Technology, Andhra Pradesh, India; and Dr. M.V Nageshwara Rao, HOD, Department of ECE, GMR Institute of Technology, Andhra Pradesh, India, for their continuous substantial co-operation, motivation, and support; without them, this work would not have been possible.

My special thanks to all my colleagues for helping me in reaching the logical conclusion of my idea in the form of this textbook.

I am indebted to my dear students as the interaction with them helped me a lot in understanding their needs.

Expressing gratitude publicly to wife is almost missing in Hindu society but I would like to extend my sincere thanks and appreciation to my wife Dr. Pratibha Misra for sparing me from day-to-day work for completing this work. I am also thankful to my son Ishan and daughter Saundarya for providing me special ideas about this book.

I acknowledge the support from Arduino for using their product images and data to demonstrate and explain the working of the systems. I thank Taylor & Francis/CRC Press team for encouraging and supporting me continuously to complete my idea about this book.

Utmost care is taken for the circuits and programs mentioned in the text. All the programs are tested on real hardware but in case of any mistake, I extend my sincere apologies. Any suggestions to improve the contents of this book are always welcome and will be appreciated and acknowledged.

I am also very much thankful to all who are directly or indirectly involved in the accomplishment of this task.

Dr. Yogesh Misra

Author

Dr. Yogesh Misra, BE (Electronics), ME (Electronics & Communication Engineering), and Ph.D., has a passion for teaching. He is having 24 years of teaching and industrial experience. He is currently working as a Professor in the NBA accredited Electronics & Communication Engineering Department of GMR Institute of Technology, Rajam (Autonomous NAAC "A" Grade Institution) affiliated to JNTU, Kakinada, Andhra Pradesh, India. All the seven programs offered by GMR Institute of Technology are accredited by NBA. GMR Institute of Technology is ranked in the band of 200 to 250 at the national level by NIRF, MHRD, New Delhi in June 2020.

Dr. Misra switched to academics in the year 2003 and worked in KL University, Andhra Pradesh, Mody University, Rajasthan, and BRCM College of Engineering and Technology, Haryana. He worked in Automatic Electronic Control System (a sugar mill automation company) as an Engineer (Installation & Commissioning) for many years.

He has conducted a number of workshops for the students/faculties on topics Interfacing and Programming of Arduino, Circuit Designing using PSPICE OrCAD, PCB Designing & Fabrication, Application of VHDL in System Design, Microprocessor and Microcontroller Programming, VLSI Embedded Computing.

He has delivered expert lectures in different institutes on topics Interfacing and Programming of Arduino, Embedded System Design, IC Manufacturing, FPGA Architecture and Application, Fuzzy Logic and its Applications and Motivating youth for Entrepreneurship. He has delivered lectures in AICTE-ISTE Refresher programs and UGC sponsored seminars.

He has published more than 30 research papers and the majority of his papers are in the field of VLSI design, soft computing, and embedded systems.

He has authored two books titled *Digital System Design using VHDL* with the Dhanpat Rai Publication Co. in 2006 and *Application of fuzzy logic in sugar mill* with the Lambert Academic Publishing in 2017.

Dr. Misra declared one of the toppers in the "Microprocessors and Microcontrollers" Course conducted by NPTEL, IIT Kharagpur.

He has made the following popular video lecture series:

1. Video Lecture Series on "System Design using Verilog" available online with certification at https://www.udemy.com/course/system-design-using-verilog/
2. Video Lectures on "Microprocessors and Microcontrollers" available online at https://www.youtube.com/c/dryogeshmisraprofessor

Abbreviations Used in This Book

V	Volt
mV	Millivolt
A	Ampere
μA	Microampere
mA	Milliampere
ms	Millisecond
μs	Microsecond
MΩ	Mega Ohm
ppm	Parts per million
cm	Centimeter

1 Introduction to Arduino UNO Board

LEARNING OUTCOMES

After completing this chapter, learners will be able to:

1. Understand various hardware features of the Arduino UNO board.
2. Understand various features of Integrated Development Environment (IDE) used for the development of software.
3. Understand how to download and use the Arduino IDE for the development of software.
4. Understand pin configuration and features of ATmega 328 microcontroller, which is the Arduino UNO board's main brain.
5. Understand Serial Peripheral Interface (SPI) and Inter-Integrated Circuit (I2C) serial communication protocols.

1.1 FEATURES OF ARDUINO UNO BOARD

Arduino is open-source hardware that can be used to develop embedded systems with open-source software. Arduino has gained massive popularity among students and hobbyists for making a working model. The reasons behind the popularity of Arduino are its low cost, availability of software, and easy-to-interface possibility. This book has used the Arduino UNO board's code because it is the most popular board in the Arduino family. The Arduino UNO is a microcontroller-based board having an ATmega328 microcontroller from ATmega (now MicrochipTM). Most of the Arduino boards have the majority of components, as shown in Figure 1.1. The input voltage range required for the Arduino UNO board's operation is 6–20 V, but the recommended input voltage range is 7–12 V. If the input voltage is less than 7 V, the digital output pins may supply less than 5 V, and the board may be unstable. The output current from each pin of Arduino UNO is 40 mA.

Power USB (Label 1): Arduino board can be powered on by connecting it to the computer at the USB socket using USB cable as shown in Label 1.

Power Connector (Label 2): Arduino board can be powered on by connecting it to 220 V AC by an adapter at Power Connector as shown as Label 2.

Voltage Regulator (Label 3): The voltage regulator shown in Label 3 is used to regulate the Arduino board's DC voltage.

Crystal Oscillator (Label 4): A crystal oscillator of 16 MHz is used in the Arduino UNO board to synchronize the microcontroller's various operations of ATmega 328 and Arduino UNO board.

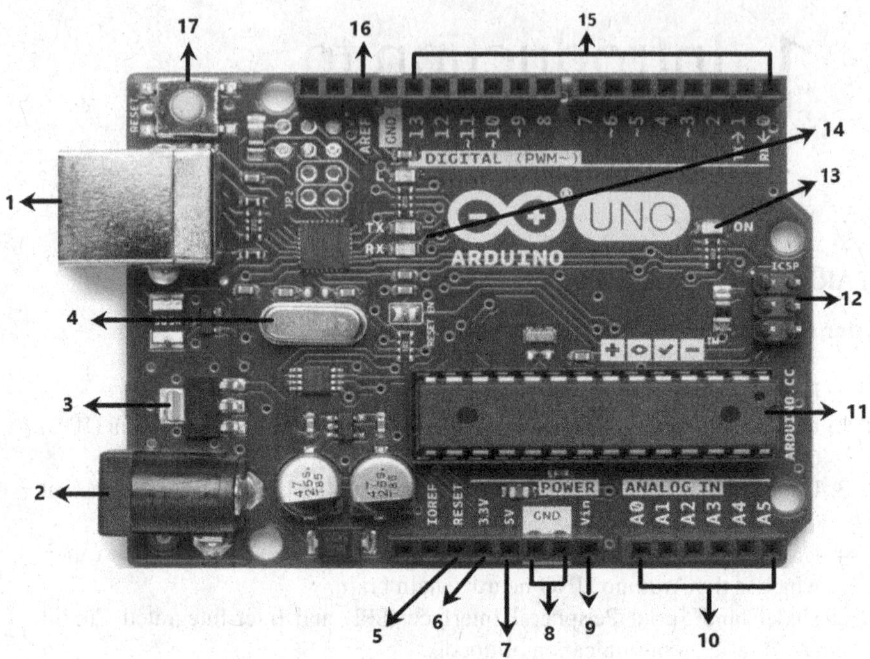

FIGURE 1.1 Arduino UNO board. (Courtesy Arduino.)

Reset (Label 5): Sometimes, a program may be stuck at some instruction. In such cases, we have to reset our Arduino board so that the program's execution should start from the beginning. If a 0 V is applied at the reset pin as shown in Label 5, the Arduino board will reset. An alternate way to reset the Arduino board is by pressing the reset button, as shown in Label 17.

3.3 V (Label 6): The 3.3V output is available at the pin as shown in Label 6.

5 V (Label 7): The 5V output is available at the pin as shown in Label 7.

GND (Label 8): In the Arduino board, three GND (ground) pins are available. Users can use any available ground pin while building their circuit. The two GND (ground) pins are available at the pins, as shown in Label 8. The third GND (ground) pin is available at the right-side pin, as shown in Label 16.

Vin (Label 9): Arduino board can be powered on by connecting a DC voltage in the range of 7–20 V at Vin pin, as shown in Label 9.

Analog Input (Label 10): The Arduino UNO board contains six analog pins named A0, A1, A2, A3, A4, and A5. Internally, these analog pins are connected to a six-channel 10-bit analog-to-digital converter. The allowable analog input voltage range at each analog input pin is 0–5 V. Since each analog input pin is connected to a 10-bit analog-to-digital converter, 0–5 V is divided into 1,024 steps ranging from 0 to 1,023. The six analog pins are shown in Label 10.

ATmega328 (Label 11): The ATmega328 microcontroller is used to perform various arithmetic and logical operations. The ATmega328 microcontroller is manufactured by Atmel Company and shown in Label 11.

ICSP (Label 12): It is a small programming header for the Arduino consisting of MOSI (Master Output Slave Input), MISO (Master Input Slave Output), SCLK (System Clock), RESET, VCC, and GND.

Power LED (Label 13): When the Arduino board is power on, the power LED turns on. The Power LED is shown in Label 13.

Tx and Rx LEDs (Label 14): When the Arduino board transmits data serially from the Arduino board through Pin 1 of the Arduino board, Tx (Transmit) LED blinks.

When the Arduino board receives data serially through Pin 0 of the Arduino board, Rx (Receive) LED blinks. The Tx and Rx LEDs are shown in Label 14.

Digital I/O (Label 15): The Arduino UNO board has 14 digital I/O (input/output) pins. Each of these pins can be configured as an input or output pin by using the *pinMode* function.[1] If the pin is configured as an input pin, then the digital signal can be applied on the pin, and if the pin is configured as an output pin, then the digital signal will output on the pin. The digital I/O pins are shown in Label 15.[2]

There are six pins in the Arduino UNO board, namely, 3, 5, 6, 9, 10, and 11, with Pulse Width Modulation (PWM) capabilities. The six PWM pins are labeled as "~".

Serial data transfer uses Pins 0 and 1. Pin 0 of the Arduino board can be used to receive data serially, and Pin 1 of the Arduino board can be used to transmit data serially. If Pins 0 and 1 are not used in serial data transfer, they can be used as digital I/O pins.

AREF (Label 16): AREF (Analog Reference) pin is sometimes used with analog input pins A0–A5. If we wish to set an external reference voltage for A0–A5 analog pins other than 0 to 5 V, we need an AREF pin. The AREF pin is shown in Label 16.

1.2 ARDUINO IDE SOFTWARE

The IDE is the software used for the development of programs. The IDE for high-level language (Embedded C)-based development flow is shown in Figure 1.2. The function of various components of IDE is shown in the following.

1.2.1 Editor

High-level language program is written in the editor by using statements. The file is saved with .ino extension for Arduino and .C/.C++ for C language. Editor generates .ino/.c/.c++ file.

1.2.2 Compiler

Compiler is a program that converts high-level language program into binary language. The object file (.o) is generated by the compiler in Arduino and .obj file for the source file of C/C++. The name of the compiler in Arduino IDE is AVR-GCC. The cross-compiler used in Keil µVision IDE is C51.

1.2.3 Linker

The purpose of a linker is to link more than one object file of the same project into one object file.

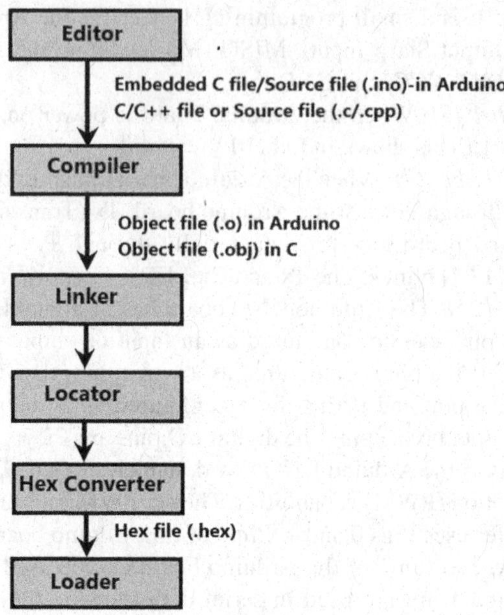

FIGURE 1.2 The process flow of software development using a high-level language (Embedded C).

1.2.4 Locator

The locator's purpose is to assign the memory address of code memory to the instructions. Locator generates Absolute Object file. The avr-ld utility is used for Linker/Locator in Arduino IDE. The Linker/Locator used in Keil µVision IDE is BL51.

1.2.5 Hex Converter

This software converts the Absolute Object file into a hexadecimal file with .hex extension. Avr-object copy utility is used to convert object files to the hex file in Arduino IDE.

1.2.6 Loader

Loader software is used to load the hexadecimal file into the target controller. The avr-dude utility is used for Linker/Locator in Arduino IDE.

1.3 ARDUINO IDE DOWNLOAD

The steps to download the IDE for the development of programs are as follows:

Step 1: Go to www.arduino.cc website.
Step 2: Go to the "DOWNLOADS" option of the "SOFTWARE" menu.
Step 3: We shall be directed to the "Download the Arduino IDE" option.
Step 4: Click the "Windows Installer for Windows XP and up" option.

Step 5: We shall be directed to the "Contribute to the Arduino Software" page.
Step 6: We shall get the "Just Download" and "Contribute & Download" option.
Step 7: After installing the .exe file, an icon of Arduino will be created on the desktop.

1.4 WORKING WITH ARDUINO IDE

The Arduino programs are called a sketch. A sketch is saved with extension .ino. The steps to work with Arduino IDE for the development of programs are as follows:

Step 1: Open the editor window to write a program in Arduino IDE. The screenshot of Step 1 is shown in Figure 1.3.
Step 2: Select the Arduino/Genuino UNO board from the "Board" option of "Tools". The screenshot of Step 2 is shown in Figure 1.4.
Step 3: Follow the procedure to know the serial port to which Arduino is mapped (Figure 1.5):

i. Right-click on My Computer.
ii. Select the Manage option.
iii. In the pop-up screen for Computer Management, select the Device Manager.
iv. Expand the Ports item; the Arduino UNO will appear as one of the drop-down items.

Step 4: In the Arduino IDE, select the "port" in the "Tool" option, and check "COM6 (Arduino/Genuino Uno)" as shown in Figure 1.6.
Step 5: The "verify", "upload", and "serial monitor" buttons on the sketch are shown in Figure 1.7. The functions of "verify", "upload", and "serial monitor" buttons are as follows:

i. The "verify" button is used to compile the program. If there are errors, the line numbers of the errors are shown in the bottom window. Correct the errors, and again click the "verify" button.

sketch_apr11a | Arduino 1.8.5
File Edit Sketch Tools Help

sketch_apr11a

```
void setup() {
  // put your setup code here, to run once:

}

void loop() {
  // put your main code here, to run repeatedly:

}
```

FIGURE 1.3 The screenshot for Step 1.

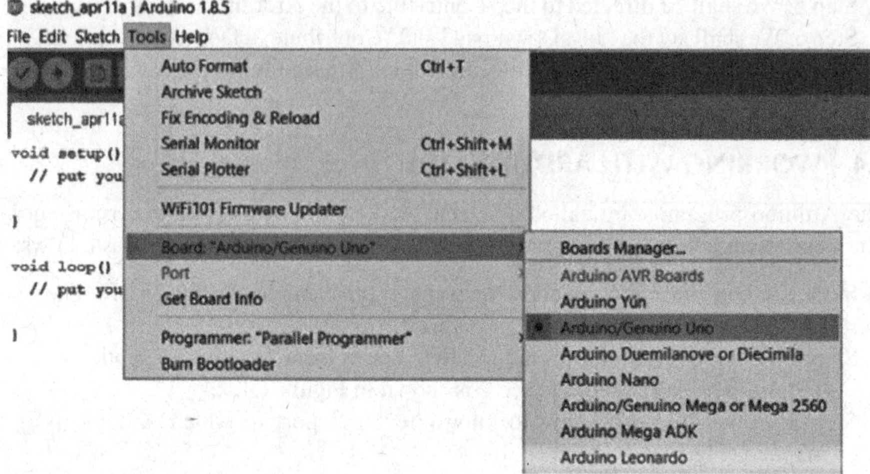

FIGURE 1.4 The screenshot for Step 2.

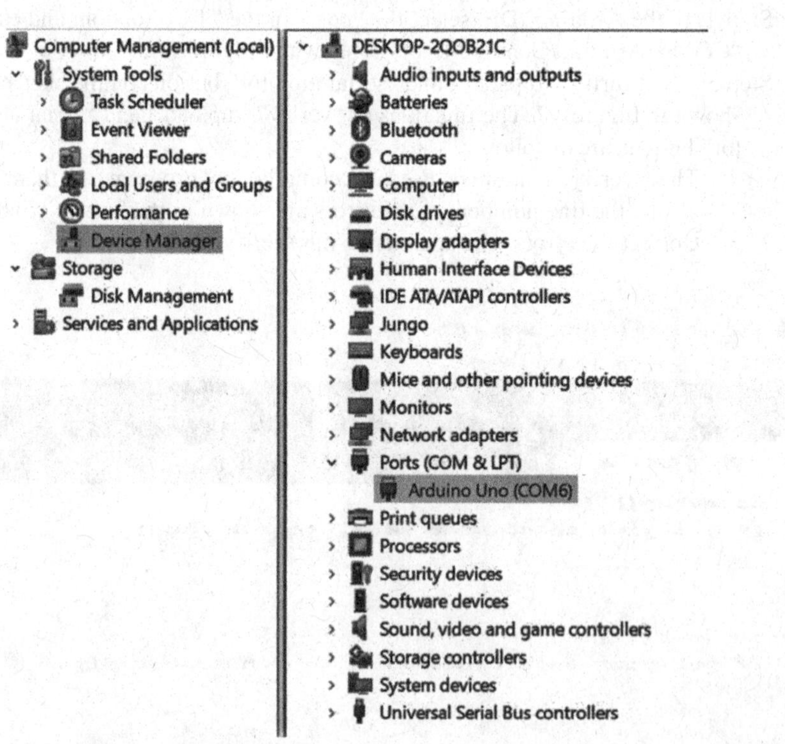

FIGURE 1.5 The screenshot for Step 3.

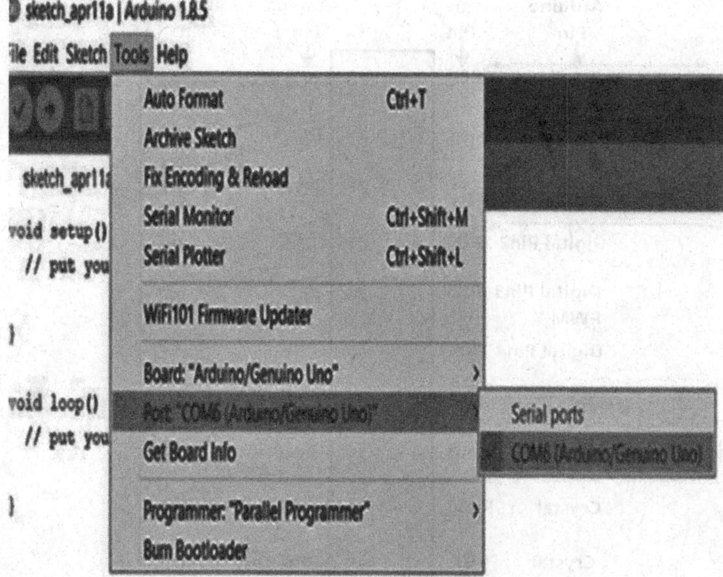

FIGURE 1.6 The screenshot for Step 4.

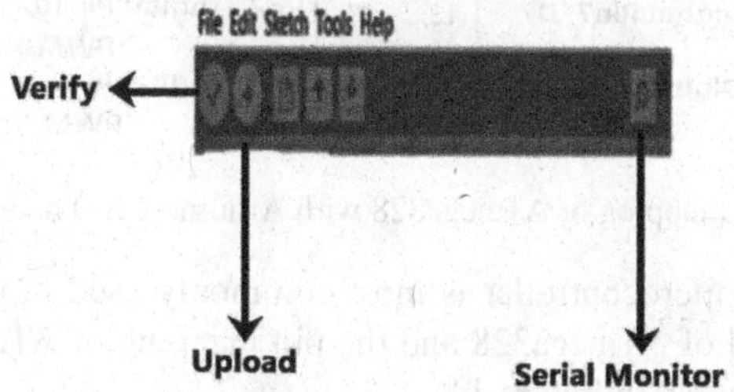

FIGURE 1.7 The screenshot for Step 5.

ii. Once the compilation is done, upload the program from the computer to the Arduino board by clicking the "upload" button. During the uploading process, the Arduino board must be connected to the computer via USB cable.

iii. To display the values, click the "serial monitor" button.

1.5 INTRODUCTION TO ATMEGA 328

The ATmega328 microcontroller was manufactured by Atmel Company; later, Atmel was acquired by Microchip Technology in 2016. It is an 8-bit RISC (Reduces Instruction Set Computer) architecture-based microcontroller. ATmega328 is a 28-pin IC having Harvard-based architecture.

Arduino Pin	ATmega Pin			ATmega Pin	Arduino Pin
Reset	PC6	1	28	PC5	A5
Digital Pin0 (RX)	PD0	2	27	PC4	A4
Digital Pin1 (TX)	PD1	3	26	PC3	A3
Digital Pin2	PD2	4	25	PC2	A2
Digital Pin3 PWM	PD3	5	24	PC1	A1
Digital Pin4	PD4	6	23	PC0	A0
Vcc	Vcc	7	22	GND	GND
GND	GND	8	21	Aref	Aref
Crystal	PB6	9	20	AVcc	Vcc
Crystal	PB7	10	19	PB5	Digital Pin13
Digital Pin5 PWM	PD5	11	18	PB4	Digital Pin12
Digital Pin6 PWM	PD6	12	17	PB3	Digital Pin11 PWM
Digital Pin7	D7	13	16	PB2	Digital Pin10 PWM
Digital Pin8	PB0	14	15	PB1	Digital Pin9 PWM

FIGURE 1.8 The pin mapping of ATmega328 with Arduino UNO board.

The ATmega328 microcontroller is most commonly used in the Arduino UNO board. The pin detail of ATmega328 and the pin mapping of ATmega 328 with the Arduino board are shown in Figure 1.8.

The ATmega328 has three ports, namely, Port B, Port C, and Port D. The three ports of ATmega328 comprise 23 pins. Port B is an 8-bit port, namely, PB0–PB7; Port C is a 7-bit port, namely, PC0–PC6; and Port D is an 8-bit port, namely, PD0–PD7. There are two ground pins available at Pins 8 and 22, one Vcc pin at Pin 7, Aref (analog reference) at Pin 21, and AVcc (analog Vcc) at Pin 20. The PB6 and PB7 pins of ATmega328 are used to connect external crystal.

Features of ATmega328:

i. It is based on RISC architecture.
ii. It is an 8-bit microcontroller.
iii. It has 131 instructions. The 95% instructions need one clock cycle for execution, 3% instructions need two clock cycles for execution, and 2% instructions need three clock cycles for execution.
iv. Its maximum clock speed is 16 MHz.
v. It has 32-KB flash code memory.

vi. It has 1,024 bytes (1 KB) EEPROM (Electrically Erasable Programmable Read-Only Memory).
vii. It has 2-KB SRAM (Static Read and Write Memory).
viii. It has two 8-bit timer/counter and one 16-bit timer/counter.
ix. It supports a two-wire I2C serial communication protocol.
x. It supports SPI serial communication protocol.
xi. It has an in-built 10-bit analog-to-digital converter.

1.6 SERIAL PERIPHERAL INTERFACE (SPI) COMMUNICATION PROTOCOL

The SPI is a synchronous serial communication protocol since master and slave both share the same clock.[3] It is suitable for interfacing microcontrollers with sensors/display/memory devices. Master is controlling devices like microcontrollers, and slaves take instructions from the master. The signals of SPI protocols are MOSI, MISO, SCLK, and SS/CS (Slave Select/Chip Select). The SS/CS signal is active low.

Slave Select: Master can select the slave with which it wants to communicate by setting the SS/CS line at a low level. In idle conditions, the SS/CS line is kept at a high-voltage level. The master sends data to the slave bit by bit, in a serial form through the MOSI line, and the slave receives data sent from the master at the MOSI pin.[4] Data sent by the master to slave is usually with MSB (most significant bit) first. The slave can also send back data to the master through the MISO pin.[1] Data sent by the slave to master is usually with LSB (least significant bit) first. There is a single master and single slave in the simple SPI configuration, as shown in Figure 1.9. Multiple SS/CS lines may be available in the master, allowing multiple slaves to be connected in parallel, as shown in Figure 1.10. If only one SS/Cs line is available in master, then multiple slaves can be connected in the Daisy-chaining method as shown in Figure 1.11.

1.6.1 Steps of SPI Data Transmission – The Steps for Data Transmission in SPI Protocols are as Follows:

i. Master outputs clock signal through the SCLK pin.
ii. Master switches the SS/CS pin to a low-voltage state, which activates the slave.
iii. The master sends data through the MOSI pin.

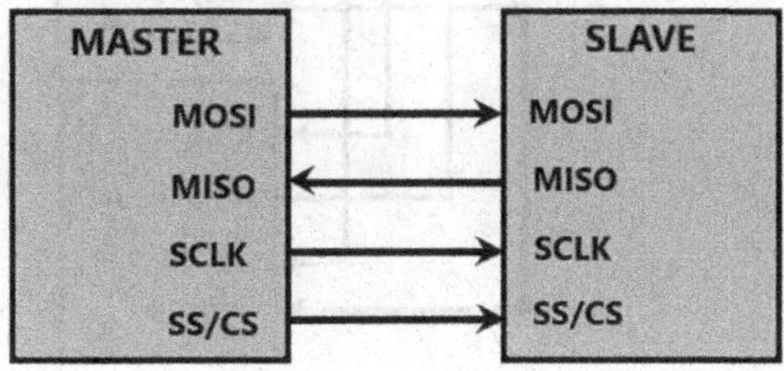

FIGURE 1.9 The simplest SPI configuration.

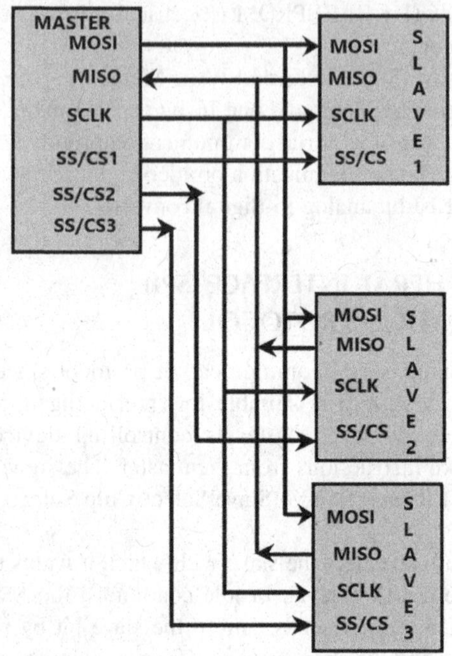

FIGURE 1.10 Multiple slaves connected to master having multiple SS/CS pins.

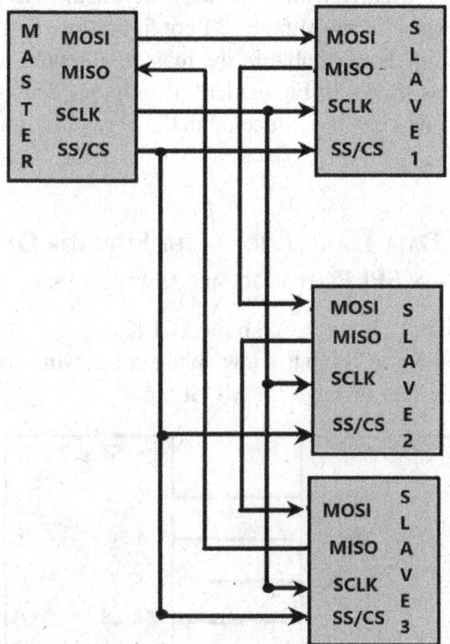

FIGURE 1.11 Multiple slaves connected to master having single SS/CS pins.

1.6.2 Advantages of SPI Data Transmission – The Advantages of Data Transmission in SPI Protocols are as Follows:

i. No start and stop bit required.
ii. No complicated slave addressing system.
iii. Higher data transfer rate than the I2C protocol (almost twice).
iv. Separate MOSI and MISO pins to transmit and receive data simultaneously.

1.6.3 Disadvantages of SPI Data Transmission – The Disadvantages of Data Transmission in SPI Protocols are as Follows:

i. It uses four wires, whereas I2C and UART (Universal Asynchronous Receiver Transmitter) use two wires only.
ii. No acknowledgment by the receiver that data has been received successfully, whereas I2C has this.
iii. It allows only a single master.

1.7 INTER-INTEGRATED CIRCUIT (I2C) COMMUNICATION PROTOCOL

The I2C serial communication protocol was invented by Phillips in 1982. The I2C protocol can connect multiple slaves to a single master or multiple slaves to multiple masters. The I2C protocol can be used when two microcontrollers send data to the same memory or send data to the same LCD, and there can be many more applications. The I2C serial communication protocol has two signals: Serial Data (SDA) and Serial Clock (SCL). The master and slave send and receive data through the SDA line, and SCL is the clock signal line. The master always controls the clock signal. The single master and single slave connections of the I2C protocol are shown in Figure 1.12, and the single master and multiple slave connections of the I2C protocol are shown in Figure 1.13.

The message frame of the I2C serial communication protocol is shown in Figure 1.14.[1] The various fields of the frame are as follows:

Start Condition: The SDA line switches from a high-voltage level to a low-voltage level before the SCL line switches from high to low.[1]

Stop Condition: The SDA line switches from a low-voltage level to a high-voltage level after the SCL line switches from low to high.[1]

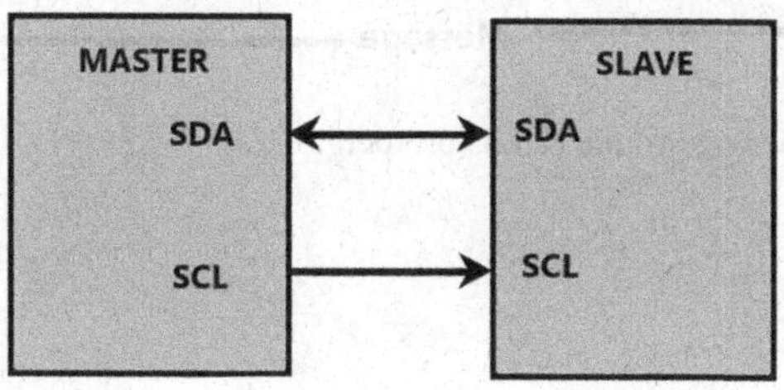

FIGURE 1.12 Single master and slave connection in I2C protocol.

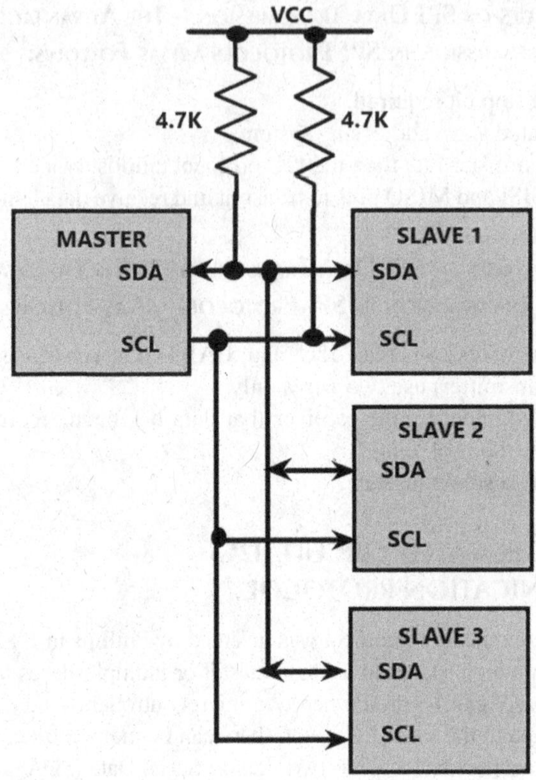

FIGURE 1.13 The single master and multiple slaves connection in I2C protocol.

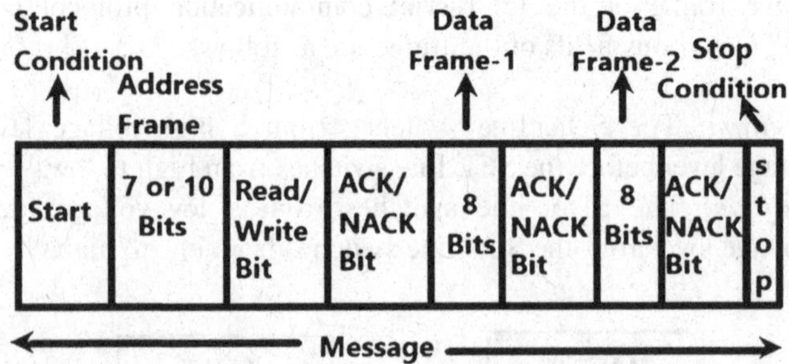

FIGURE 1.14 The message frame of I2C protocol.

Address Frame: A 7 or 10-bit sequence unique to each slave identifies the slave when the master wants to talk to it.[1]

Read/Write Bit: A single bit specifying whether the master is sending data to the slave (low-voltage level) or requesting data from it (high-voltage level).[1]

ACK/NACK Bit: Each frame in a message is followed by an acknowledge/-no-acknowledge bit. If an address frame or data frame was successfully received, an ACK bit is returned to the sender from the receiving device.[1]

Check Yourself

1. What does IDE stand for?
2. What is the name given to the Arduino program?
3. Arduino sketch has two functions. What are those?
4. How many analog input pins are used in Arduino UNO?
5. How many digital pins are used in Arduino UNO?
6. How many PWM pins are used in Arduino UNO?
7. On-board LED is connected to which pin?
8. What frequency crystal is connected to the Arduino UNO board?
9. ATmega328 is based on RISC-based or CISC-based architecture?
10. ATmega328 is an 8-bit or 16-bit microcontroller?
11. The ATmega328 has .... Flash memory, ...EEPROM and ...... SRAM.
12. The ATmega328 is a .... Pin microcontroller.
13. Which type of signal is to be applied to reset pin to reset the Arduino UNO board?
14. Which two pins of the Arduino board are used for I2C protocol and for which I2C signal?
15. Which four pins of the Arduino board are used for SPI protocol and for which SPI signal?
16. Which of the following are present in the Arduino UNO board?
    a. 64 Kbytes of programmable flash memory
    b. ATmega328 microcontroller
    c. 2 Kbytes of SRAM
    d. 16 MHz clock

2 Arduino Programming Constructs

LEARNING OUTCOMES

After completing this chapter, learners will be able to:

1. Understand the structure of the Arduino program.
2. Understand the syntax of creating a variable and constant.
3. Understand various functions that can be used while writing the Arduino program.
4. Understand the application and uses of various functions in the Arduino program.
5. Understand the application of various predefined operators of the Arduino language.

2.1 STRUCTURE OF ARDUINO PROGRAMMING

Each Arduino program must have two functions as shown below:

```
void setup()
{
}
void loop()
{
}
```

2.1.1 *Setup()*

The `setup()` function is called when the Arduino program starts. It will run only once, after each power-up or reset of the Arduino board. `Setup()` function is used to initialize variables, pin modes, libraries, serial communication, etc.

2.1.2 *Loop()*

The `loop()` function executes an infinite number of times until the Arduino board is getting power. It is used to control various devices using the Arduino board.

2.1.3 Variables

All variables have to be declared before they are used. An object of the variable class holds a single value of a specified type. However, different values can be assigned to the variable by using variable assignment statements. Declaring a variable means defining its type, and optionally, setting an initial value (initializing the variable).

Syntax of variable object declaration:

```
type-name name-of-object = initial-value;
```

Example:

```
int LED = 13;
```

In the above statement, LED has declared a variable that holds the integer-type value, and its initial value is 13. The value of the LED variable can be changed later on.

2.1.4 Constant

All constants have to be declared before they are used. An object of the constant class holds a single value of a specified type. The value of the constant cannot be changed in the program. Declaring a constant means defining its type, and optionally, setting an initial value (initializing the constant). Syntax of variable object declaration:

```
const type-name name-of-object = initial-value;
```

Example:

```
const int MY_LED = 13;
```

In the above statement, MY_LED is declared a constant that holds the value of the integer type, and its initial value is 13. The value of the MY_LED constant cannot be changed later on.

2.1.5 Integer

The values of integer types fall within the specified range of integers. The values of integer type at least cover a range from (−32,368 to +32,367). The values of integer types are called integer literals.

2.2 FUNCTION

A function is a name given to a group of code that performs some specific work.

2.2.1 *pinMode(pin, mode)*

The `pinMode` function is used to configure the specified pin as input or output. The `pinMode` function has two arguments, i.e., pin and mode. The argument pin is used to specify the pin whose mode we wish to set. The argument mode can be assigned either as input or as output by writing INPUT or OUTPUT, respectively.

Example 2.1

A LED is connected to Pin 13 of Arduino UNO. Write a program to declare Pin 13 as an output pin.

Solution:

A program (Method I) to declare Pin 13 as an output pin is shown in Figure 2.1.

int LED = 13;	*statement (1)*
void setup()	
{	
pinMode(LED,OUTPUT);	*statement (2)*
}	

FIGURE 2.1 An Arduino program to assign the name "LED" to Pin 13 of the Arduino board and initialize it as an output pin.

Description of Program shown in Figure 2.1:

In the statement (1), LED is declared a variable that holds the value of integer type, and its value is assigned as 13.
Inside `setup()` in statement (2), the `pinMode` function declares the LED variable as an output pin. Since Pin 13 is assigned to variable LED in statement (1), due to statement (2), Pin 13 is initialized as an output pin.

An alternate program to declare Pin 13 as the output pin is shown in Figure 2.2.

void setup()	
{	
pinMode(13,OUTPUT);	*statement (1)*
}	

FIGURE 2.2 An Arduino to initialize Pin 13 of Arduino board as an output pin without assigning it any name.

Description of Program as shown in Figure 2.2:

Inside `setup()` in the statement (1), the `pinMode` function is used to initialize Pin 13 as the output pin. Since we are using pin number directly in the `pinMode` function, there is no need for statement (1) of the Method I program.

2.2.2 DIGITALWRITE(PIN, VALUE)

The `digitalWrite` function is used to set the pin configured as an output pin using `pinMode` function to either a HIGH or a LOW state. The `digitalWrite` function has two arguments, i.e., pin and mode. The argument pin is used to specify the pin whose mode we wish to set.

Example 2.2

The anode of LED is connected to Pin 13 of Arduino UNO, and the cathode of LED is connected to the ground. Write a program to on the LED.

Solution:

A program to on the LED connected to Pin 13 is shown in Figure 2.3.

int LED = 13;	*statement (1)*
void setup()	
{	
pinMode(LED,OUTPUT);	*statement (2)*
}	
void loop()	
{	
digitalWrite(LED,HIGH);	*statement (3)*
}	

FIGURE 2.3 An Arduino to turn on the LED connected to Pin 13 of Arduino board.

Description of Program:

The statement (1) LED is declared a variable that holds the value of integer type, and its value is assigned as 13. By using statement (1), we wish to give the name LED to Pin 13.

Inside `setup()` in the statement (2), the `pinMode` function declares Pin 13 as an output pin. Since Pin 13 is assigned to variable LED in statement (1), in the statement (2), we have used the LED name of Pin 13 for initializing it as an output pin.

Inside `loop()`, the `digitalWrite` function in the statement (3) is used to set Pin 13 to HIGH. Due to statement (3), Pin 13 will be set at 5 V. Since the anode and cathode of LED are connected to Pin 13 and GND (ground) of Arduino UNO board, respectively, after the execution of `the digitalWrite(LED, HIGH)` statement, the LED will on.

2.2.3 `DIGITALREAD(PIN)`

The `digitalRead(pin)` function reads the digital value from the specified pin, which is given in the argument of the `digitalRead` function. This function returns the value either high (5 V/logic 1) or low (0 V/logic 0).

Example 2.3

Declare Pin 13 of Arduino UNO as an input pin, and read the digital value available at this pin.

Solution:

A program to declare Pin 13 of Arduino UNO as an input pin and read the digital value available at this pin is shown in Figure 2.4.

int SWITCH = 13;	*statement (1)*
void setup()	
{	
pinMode(SWITCH,INPUT);	*statement (2)*
}	
void loop()	
{	
int VALUE = 0;	*statement (3)*
VALUE = digitalRead(SWITCH);	*statement (4)*
}	

FIGURE 2.4 An Arduino to initialize Pin 13 of Arduino board as an input pin and read its value.

Description of Program:

In statement (1), SWITCH has declared a variable that holds the value of integer type, and its value is assigned as 13. By using statement (1), we wish to give the name SWITCH to Pin 13.

Inside `setup()` in the statement (2), the `pinMode` function declares Pin 13 as the input pin. Since Pin 13 is assigned to variable SWITCH in statement (1), in the statement (2), we have used the SWITCH name of Pin 13 for initializing it as an input pin.

Inside `loop()`, the statement (3) is used to declare SWITCH as a variable that holds the value of integer type, and its value is assigned as 0.

Inside `loop()`, the `digitalRead` function in the statement (4) is used to read the digital value of input Pin 13 (named SWITCH). The digital value read from Pin 13 is assigned to the variable VALUE.

2.2.4 ANALOGREAD(PIN)

The *`analogRead(pin)`* function reads the analog value from the specified analog pin as given in the analogRead function's argument. The Arduino board contains six analog pins named A0, A1, A2, A3, A4, and A5. Internally, these analog pins are connected to a six-channel 10-bit analog-to-digital converter. The allowable analog input voltage range at each analog input pin is 0–5 V. Since each analog input pin is connected to a 10-bit analog-to-digital converter, 0–5 V is divided into 1,024 steps. So there are 1,024 steps ranging from 0 to 1,023 steps. The analogRead function returns an integer value in the range of 0–1,023.

* *Please refer to Section 3.6 of Chapter 3 for analog-to-digital converter.*

Example 2.4

Declare the Pin A0 of Arduino UNO as an input pin, and read the analog value available at this pin.

Solution:

A program to declare the Pin A0 of Arduino UNO as an input pin and read the analog value available at this pin is shown in Figure 2.5.

int SIGNAL = A0;	*statement (1)*
void setup()	
{	
pinMode(SIGNAL,INPUT);	*statement (2)*
}	
void loop()	
{	
int myValue = 0;	*statement (3)*
myValue = analogRead(SIGNAL);	*statement (4)*
}	

FIGURE 2.5 An Arduino to initialize analog Pin A0 of Arduino board as an input pin and read its value.

Description of Program:

In the statement (1), SIGNAL has declared a variable that holds the value of integer type, and its value is assigned analog Pin A0.

Inside *setup()* in the statement (2), the pinMode function declares the A0 analog pin as an input pin.

Inside *loop()*, the statement (3) is used to declare myValue as a variable that holds the value of integer type, and its value is assigned as 0.

Inside *loop()*, the analogRead function in the statement (4) is used to read the input Pin A0's analog value and named as SIGNAL. The analog value read from Pin A0 is assigned to variable myValue. The value assigned to myValue will be in the range from 0 to 1,023.

2.2.5 *ANALOGWRITE(PIN, VALUE)*

The *analogWrite* function is used to write an analog value to pulse width modulation (PWM) pins. There are six pins in the Arduino UNO board, namely, 3, 5, 6, 9, 10, and 11, with PWM capabilities. The *analogWrite* function is not associated with any of the six analog input pins A0–A5. The *analogWrite* function has two arguments, i.e., pin and value. The argument pin is used to specify the PWM pin, and the argument value can have a value from 0 to 255. The *analogWrite* writes the value from 0 to 255 in the PWM pin. The *analogWrite* function can light an LED at varying brightness or drive a motor at various speeds. Readers can find programs related to PWM concepts in chapter 4.

2.2.6 *DELAY(VALUE)*

The *delay* is a function that generates a delay for a specified amount of time. The *delay* function has one argument, and it generates a delay equal to the value of the argument in millisecond (ms). The generation of delay means the microcontroller will be engaged in executing a few sets of instructions for the specified amount of time. The microcontroller will execute the next instruction written after the *delay* function only after the specified time, as mentioned in the delay function's argument.

2.2.7 *FOR* LOOP

The *for* loop is used to repeat a group of statements enclosed in curly brackets. The *for* statement is useful for executing a group of statements unless a condition meets. The syntax of *for* loop statement is as follows[4]:

```
for (initialization; condition; increment)
{
//statement(s);
}
```

The initialization happens first and exactly once. Each time through the *loop*, the condition is tested; if it's true, the statement block and the increment are executed, and the condition is tested again. When the condition becomes false, the *loop* ends.

2.2.8 *SERIAL.BEGIN(RATE)*

The *begin* function of the "Serial" library has one argument, *Serial.begin* function is used to initialize the baud rate for serial communication. This function is used to initiate the serial communication between the Arduino UNO board and other devices like computer, sensor, input or output device, etc.

The integer-type value is allowed in the argument of the *Serial.begin* function. For example, the *Serial.begin(9600)* will initialize the serial communication between the Arduino UNO board and other peripheral at 9,600 Baud. The 9,600 Baud signifies that the data will be communicated at the rate of 9,600 bits per second during serial communication.

When the Arduino UNO board is initialized for serial communication, Pins 0 and 1 will be used for serial data transfer. Pin 0 of the Arduino board will be used to receive data serially, and Pin 1 of the Arduino board will be used to transmit data serially. If Pins 0 and 1 are not used in serial data transfer, then these pins can be used as digital I/O pins. But if we are using Pins 0 and 1 for serial data transfer, then these pins cannot be used as digital I/O pins.

2.2.9 *SERIAL.PRINT("ARGUMENT")*

The *Serial.print* function is used to print on serial monitor. The *Serial.print* function has one argument, and it is placed between double quotes.

The "Serial.print("argument")" function will print the argument on the serial monitor, and after printing, the cursor will remain in the same line.

2.2.10 *SERIAL.PRINTLN("ARGUMENT")*

The *Serial.println* function is also used to print on serial monitor. The *Serial.println* function has one argument, and it is placed between double quotes.

The *Serial.println("argument")* function will print the argument on the serial monitor, and after printing, the cursor will go to the next line.

2.2.11 *IF* STATEMENT

The *if* statement is used to select a collection of statements to execute if a certain condition is true. The condition is Boolean; i.e., on evaluation, it returns the value "TRUE" or "FALSE". The *if* statement requires one or more of the following operators for evaluation:

x == y (x is equal to y)
x != y (x is not equal to y)
x<y (x is less than y)
x>y (x is greater than y)
x <= y (x is less than or equal to y)
x >= y (x is greater than or equal to y)

The readers must be aware of "=" and "==" operators. The single equal sign is the assigned operator. For example, *if (x = 10)* sets the value of x to 10.

The double equal sign is the comparison operator. For example, *if (x == 10)* compares the value of x with 10. This statement will be true if the value of x is 10.

The three forms of *if* statement are described in the following.

2.2.11.1 Simple *if*

In this form of *if* statement, the collection of statements are selected for execution based on a single condition.

Syntax:

```
if (condition)
{
Statements
}
```

Example 2.5

Explain a simple `if statement` with the help of an example.

Solution:

A part of a program having a simple *if* statement is shown in Figure 2.6.

void loop()	
{	
int number;	
if (number > 100)	
{	
digitalWrite(LED,HIGH);	*statement (1)*
}	
}	

FIGURE 2.6 Part of Arduino program to explain the use of simple if statement.

Description of Program:

If the variable number is greater than 100, then statement (1) will be executed; otherwise, statement (1) will be skipped.

2.2.11.2 *if/else*

In this form of `if` statement, one of the two collections of statements is selected for execution based upon the single condition.

Syntax:

```
if (condition)
{
Statements
}
else
{
Statements
}
```

Example 2.6

Explain simple `if/else` statement with the help of an example.

Solution:

A part of program having `if/else` statement is shown in Figure 2.7.

Description of Program:

If the variable number is greater than 100, then statement (1) will be executed; otherwise, statement (2) will be executed.

void loop()	
{	
int number;	
if (number > 100)	
{	
digitalWrite(LED1,HIGH);	*statement (1)*
}	
else	
{	
digitalWrite(LED2,HIGH);	*statement (2)*
}	
}	

FIGURE 2.7 Part of Arduino program to explain the use of a single "if-else statement".

2.2.11.3 Multiple `if/else`

In this form of `if` statement depending upon the condition's value, anyone's collection of statements is executed. `if` statement can have zero or more `else if` clause and an optional `else` clause.

Syntax:

```
if (condition)
{
Statements
}
else if (condition)
{
Statements
}
else if (condition)
{
Statements
}
```

Example 2.7

Explain multiple `if/else` statement with the help of an example.

Solution:

A part of program having multiple `if/else` statement is shown in Figure 2.8.

Description of Program:

If the variable number is less than 100, then statement (1) will be executed; if the number is greater than or equal to 500, then statement (2) will be executed; otherwise, statement (3) will be executed.

void loop()	
{	
int number;	
if (number < 100)	
{	
digitalWrite(LED1,HIGH);	*statement (1)*
}	
else if (number >= 500)	
{	
digitalWrite(LED2,HIGH);	*statement (2)*
}	
else	
{	
digitalWrite(LED3,HIGH);	*statement (3)*
}	
}	

FIGURE 2.8 Part of Arduino program to explain the use of multiple "if-else statement".

2.2.12 MAP FUNCTION

The map function is used to assign the value from one range to another range. The syntax of the map function is as follows:

```
map(value, lowFrom, highFrom, lowTo, highTo)
```

The "value" is the range of numbers used for mapping. The lowFrom and highFrom are the lower and higher numbers from the range of numbers taken from "value". The lowFrom values and highFrom values are mapped with lowTo and highTo values. The lowFrom number from "value" is mapped with lowTo, and the highFrom number from "value" is mapped with highTo.

For example, `map(step, 0,1023,0.0,5.0);`

Let us assume step is a variable of integer type having values in the range from 0 to 1,023. The numbers in the range (0–1,023) are mapped in between "0.0–5.0 V", where step value 0 mapped with 0.0 V and 1,023 mapped with 5.0 V.

Check Yourself

1. What does the statement `int SWITCH = 13`; indicate?
2. When writing a sketch, how do you decide which features belong in the setup function and which belong in the `loop` function?
   a. Features that need to be initialized go in `setup`
   b. Features that need to be initialized go in the `loop`
   c. Features that need to run continuously go in `setup`
   d. Features that need to run continuously go in the `loop`
3. Select the function that you can use to detect a button press on the Arduino
   a. `buttonRead()`
   b. `buttonPress()`
   c. `analogRead()`
   d. `digitalRead()`
4. `delay(1000)` results in a delay of ------------
5. The `Serial.print` and `Serial.println` perform the same operation. (True/False)
6. Identify the error in `int switch=14` statement.
7. Identify the error in `Pinmode(13,OUTPUT)` statement.

3 I/O Devices, Actuators, and Sensors

LEARNING OUTCOMES

After completing this chapter, learners will be able to:

1. Understand the working principle and function of various pins of output devices such as light-emitting diode (LED), seven-segment display, and liquid crystal display (LCD).
2. Understand the working principle and function of various pins of switch, keypad matrix, potentiometer, and analog-to-digital converter IC.
3. Understand the working principle and function of various pins of motor driver board (L293D) and relay board.
4. Understand the working principle and function of various pins of temperature sensor (LM35), humidity and temperature sensor (DHT11), light-dependent register (LDR), touch sensor, smoke detector (MQ2), rain detector (FC-07), ultrasonic sensor (HC-SR04), and moisture sensor (YL-69).
5. Understand the function of various Bluetooth module pins (HC-05) and GSM module (SIM 900A).

3.1 LIGHT-EMITTING DIODE (LED)

A LED is shown in Figure 3.1. The LEDs are available in a variety of colors like red, orange, green, etc. A LED is a two-terminal device. The longer leg is called an anode, and a smaller leg is called a cathode. The anode and cathode terminals are internally connected to p-type and n-type semiconductor materials respectively. The symbol of LED is shown in Figure 3.2.

3.1.1 LED under Forward Bias

When the anode and cathode terminals of a LED are connected to positive (+ve) and negative (–ve) terminals of a battery, respectively, the LED will become forward-biased, and current flows through it. During forward bias, the LED turns on and emits light. A LED under forward-biased condition is shown in Figure 3.3 (a).

3.1.2 LED under Reverse Bias

When the anode and cathode terminals of a LED are connected to the negative (–ve) and positive (+ve) terminals of a battery, respectively, the LED will become reverse

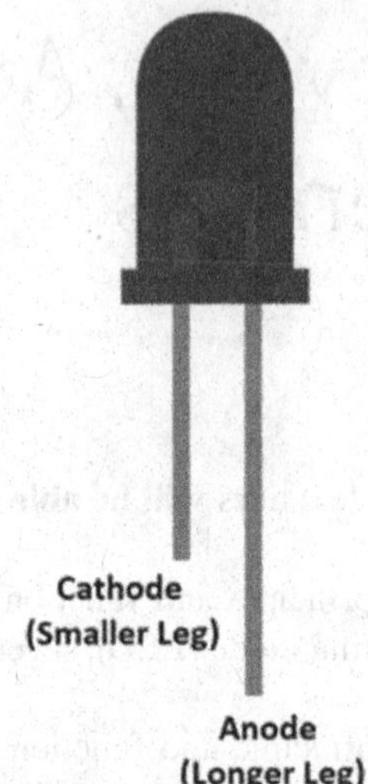

FIGURE 3.1 Light-emitting diode (LED).

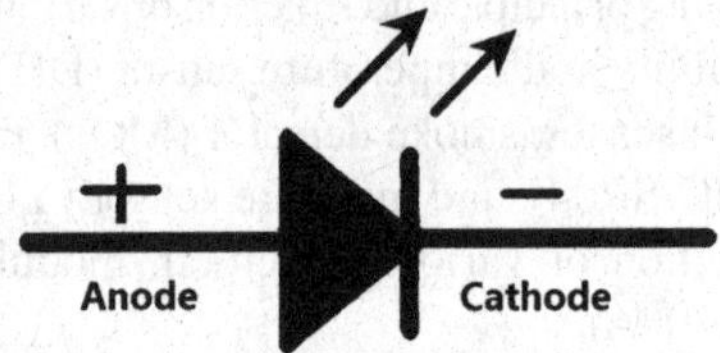

FIGURE 3.2 Symbol of light-emitting diode (LED).

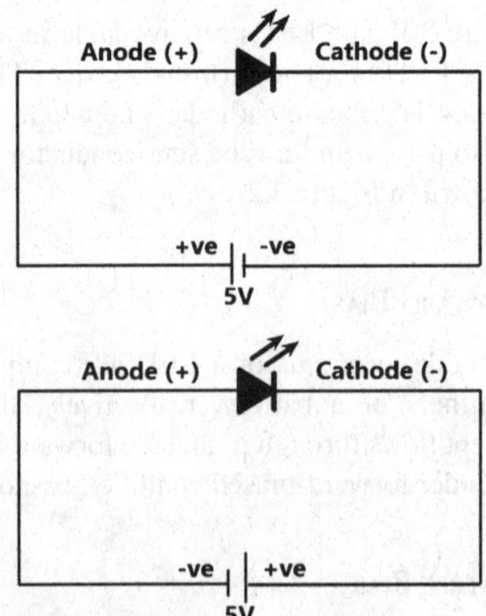

FIGURE 3.3 (a) Circuit diagram of a LED under forward bias condition. (b) Circuit diagram of a LED under reverse bias condition.

bias, and current conduction through it stops. During reverse bias, the LED turns off and does not emit light. A LED under reverse bias condition is shown in Figure 3.3 (b). The semiconductor theory is out of the scope of this book.

3.1.3 Interfacing of LED with Arduino UNO

The circuit diagram of interfacing a LED with Pin 13 of the Arduino UNO board is shown in Figure 3.4. The DC flow out from each pin of Arduino UNO is 40 mA when the particular pin is at high (5 V) level. But the LED requires a current in the range of 15–20 mA, depending upon the variety of LED. If we connect the LED directly to Pin 13 of the Arduino board, then the 40 mA DC which flows out from the pin will burn out the LED. So we have to bring down the 40 mA current of the Arduino pin to a safe level.

We know that V = IR

Here, V = Voltage at Pin 13 of Arduino board, I = Current passing through LED, and R = Resistor connected to anode of LED and Pin 13 of Arduino board.

Therefore, R = V/I

Here, V = 5 V and I = 20 mA (the current which LED can withstand).

Now, R = 5 V/15 mA

The value of R is 250 Ω.

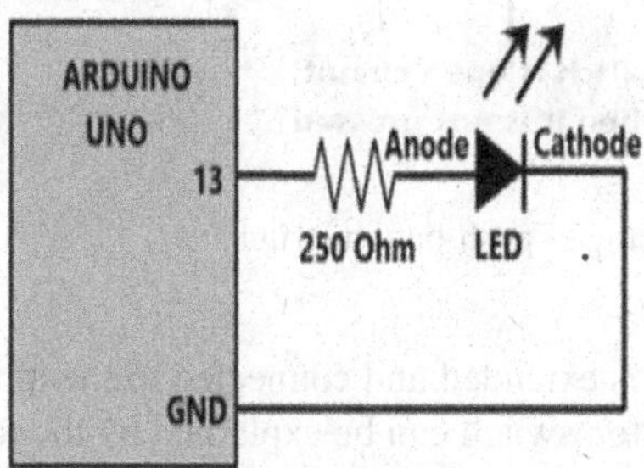

FIGURE 3.4 Circuit diagram for interfacing a LED with pin number 13 of Arduino UNO board.

** Refer to Sections 4.1 and 4.5 of Chapter 4 for programming and interfacing of the LED with Arduino UNO.*

3.2 SWITCH

A switch is a mechanical device that is used to make or break an electric circuit.

In this section, we shall discuss a push-button type of switch. It is a very popular switch. A push-button type of switch is shown in Figure 3.5. It has two terminals for connection, namely, Terminal 1 and Terminal 2, and one pushing pad. A circuit diagram depicting the interfacing of a push-button switch is shown in Figure 3.6. The Terminal 2 of the push button is connected to one end of the 1 KΩ resistor. The second end of the resistor is connected to the +5 V power supply. The Terminal 1 of the push button is connected to the GND (ground). The junction of Terminal 2 and one

FIGURE 3.5 The push-button switch.

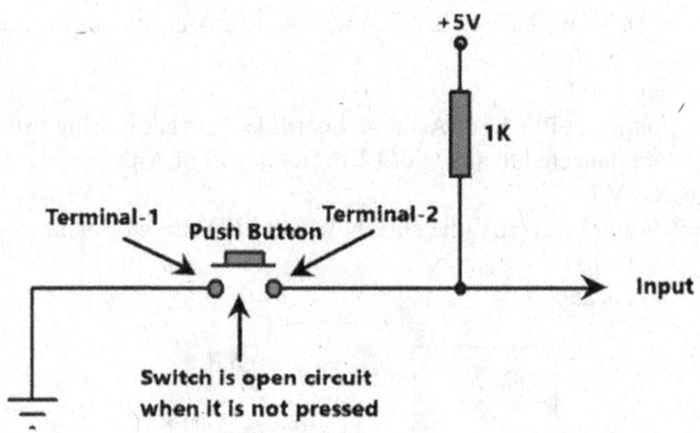

FIGURE 3.6 The interfacing of push-button switch.

end of the 1 KΩ resistor is extended and connected to the input terminal. The working principle of push-button switch can be explained by the following two cases:

Case 1: The case 1 is shown in Figure 3.6. During the initial state when the pushing pad of the switch is not pressed, Terminal 1 and Terminal 2 are open circuit and no current conducts through the switch. The input terminal always follows a low resistance path, and it has two options. Either the input terminal is connected to 5 V through 1 KΩ resistor, or it is connected to GND through an open-circuit switch. Since the resistance 1 KΩ is very small compared to the open circuit (the open circuit is considered infinite resistance), the input terminal will be connected to 5 V through resistor 1 KΩ.

Case 2: The case 2 is shown in Figure 3.7. When the pushing pad of the switch is pressed, Terminal 1 and Terminal 2 are short circuit and current conducts through the switch. The input terminal always follows a low resistance path, and it has two options. The input terminal is connected to 5 V through resistor 1 KΩ or connected to GND through a short-circuit switch. The short circuit is considered very low resistance compared to 1 KΩ; therefore, the input terminal will be connected to GND through the switch.

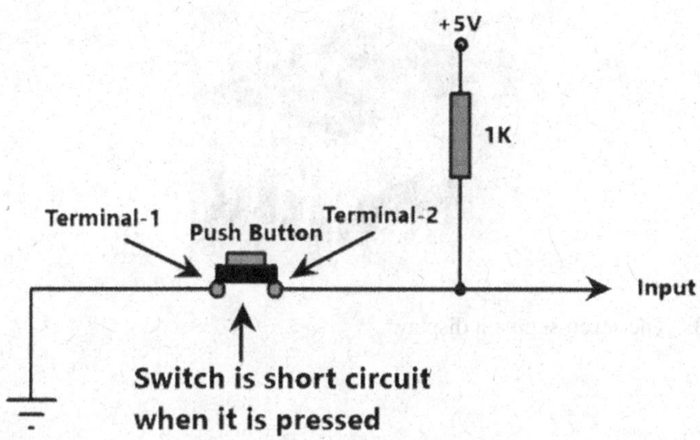

FIGURE 3.7 The interfacing circuit when push-button switch is pressed.

** Refer to Section 4.3 of Chapter 4 for programming and interfacing of the switch with Arduino UNO.*

3.3 SEVEN-SEGMENT DISPLAY

A seven-segment display is generally used to display numbers from 0 to 9. In some cases, we can also display some alphabets, but it is primarily used to display numbers. It is a low-cost solution to display information. It comes in a wide variety of colors like red, blue, orange, green, etc. A seven-segment display is shown in Figure 3.8.

The seven-segment display is available in a package of ten pins. It has seven LEDs, which are the seven segments of the seven-segment display, and these segments are visualized in the shape of number "8". These seven segments are illuminated in different sequences primarily to display numbers from 0 to 9. An additional eighth LED is also available in the seven-segment display to indicate a decimal point. The various segments and pin details of the seven-segment display are shown in Figure 3.9. The various segments are marked from a to b, and the decimal point is marked as DP. Each segment's working is based on the working principle of LED as discussed in Section 3.1. The pin details of a seven-segment display are shown in Table 3.1.

A seven-segment display is available in the market in two forms, namely, common cathode (CC) seven-segment display and common anode (CA) seven-segment display.

3.3.1 Common Cathode Seven-Segment Display (CC)

In a CC seven-segment display, all the segment's cathode terminal is joined together. To operate a CC seven-segment display, the CC terminal must be connected to the ground. To turn on, the anode terminal of the specific segment must be powered by 5 V. The internal connection of a CC seven-segment display is shown in Figure 3.10.

FIGURE 3.8 The seven-segment display.

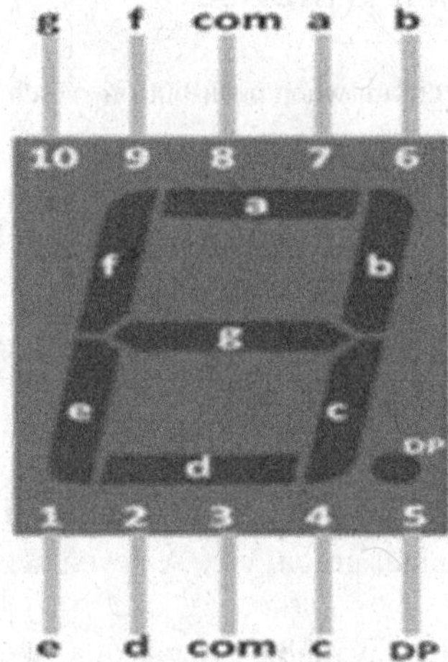

FIGURE 3.9 The various segments and pin details of seven-segment display.

TABLE 3.1
Pin Details of Seven-Segment Display

Pin Number	Description
1	It is used to control segment "e"
2	It is used to control segment "d"
3	It is connected to Vcc/ground depending upon the type of display
4	It is used to control segment "c"
5	It is used to control decimal point "DP"
6	It is used to control segment "b"
7	It is used to control segment "a"
8	It is connected to Vcc/ground depending upon the type of display
9	It is used to control segment "f"
10	It is used to control segment "g"

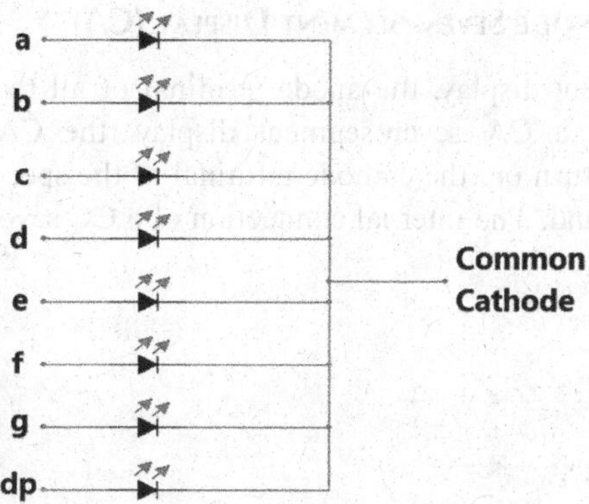

FIGURE 3.10 Common cathode seven-segment display (CC).

Example 3.1

Explain the control word to display "5" in a CC-type seven-segment display.

Solution

To display "5" in a seven-segment display, segments a, c, d, f, and g should turn on and segments b and e should turn off. The display of "5" in a CC segment display is shown in Figure 3.11. In this example, we shall also turn off the dp segment. To turn on any segment in a CC-type seven-segment display, we have to send 5 V (or binary 1) to the specific segment, and to turn it off, we have to send 0 V (or binary 0). The control word to display "5" is shown in Table 3.2.

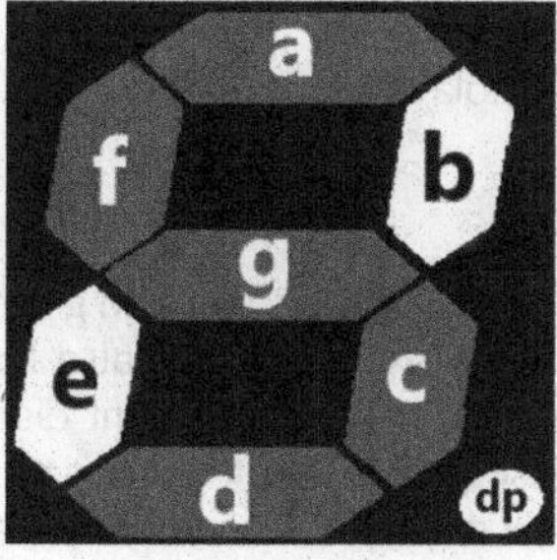

FIGURE 3.11 The display of "5" in a common cathode (CC) seven-segment display.

TABLE 3.2
Control Word to Display 5 for CC Seven-Segment Display

	Segments of Display								Number to Display
	DP	g	f	e	d	c	b	a	
Voltage level (V)	0	5	5	0	5	5	0	5	5
Binary value	0	1	1	0	1	1	0	1	

3.3.2 Common Anode Seven-Segment Display (CA)

In a CA seven-segment display, the anode terminal of all the segments is joined together. To operate a CA seven-segment display, the CA terminal must be connected to 5 V. To turn on, the cathode terminal of the specific segment must be connected to the ground. The internal connection of a CC seven-segment display is shown in Figure 3.12.

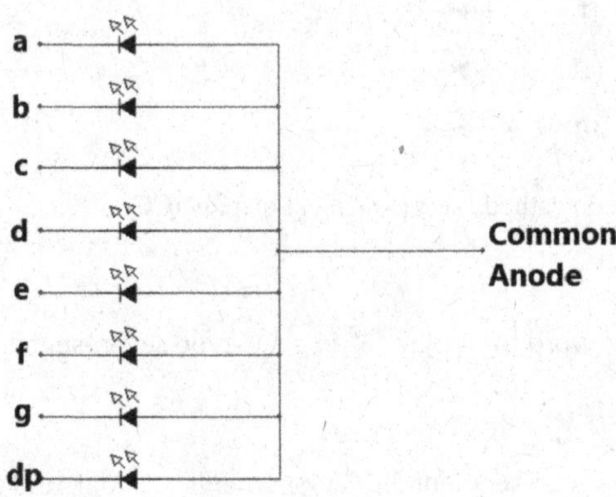

FIGURE 3.12 Common anode seven-segment display (CA).

Example 3.2

Explain the control word to display "5" in a CA-type seven-segment display.

Solution

We know that to display "5" segments a, c, d, f, and g should turn on and segments b and e should turn off. Let us assume we shall also turn off the dp segment. To turn on any segment in a CA-type seven-segment display, we have to send 0 V (or binary 0) to the specific segment, and to turn it off, we have to send 5 V (or binary 1). The control word to display "5" is shown in Table 3.3.

TABLE 3.3
Control Word to Display 5 for CA Seven-Segment Display

	Segments of Display								Number to Display
	DP	g	f	e	d	c	b	a	
Voltage level (V)	0	5	5	0	5	5	0	5	5
Binary value	1	0	0	1	0	0	1	0	

* *Refer to Section 4.4 of Chapter 4 for programming and interfacing of the seven segments with Arduino UNO.*

3.4 LCD MODULE

We have already discussed seven-segment displays. In a seven-segment display, we can display the numbers from (0 to 9), and it is possible to display a few alphabets. But in the case of LCD, we can display the numbers from (0 to 9), alphabets (capital/small), and special characters like %, @, ? etc.

Let us discuss a popular 16 by 2 LCD. In a 16×2 LCD, there are 16 columns from 0 to 15 and two rows from 0 to 1. In a 16×2 LCD, we can display 32 characters. A 16×2 LCD is shown in Figure 3.13. To display each character, we have a 5×7 pixels matrix where five is the number of columns and seven is the number of rows.

LCD is a 16-pin display device. The pin name and its function are briefly explained in Table 3.4.

Power Pins: Pin 1 and Pin 2 are the power pins. Ground and 5 V are connected to Pin 1 and Pin 2, respectively.

Contrast Control Pin: Pin 3 is used to control the contrast of the display. Generally, the output of a potentiometer is used to connect at Pin 3.

Register Select (RS) Pin: Pin 4 is used to select register. In LCD, there are two registers, viz., a command register and a data register.

The command words are to be loaded in the command register. There are many command words specified for various display controls. For example, to start displaying characters from a specific row and column can be set by loading a specific control word in the command register.

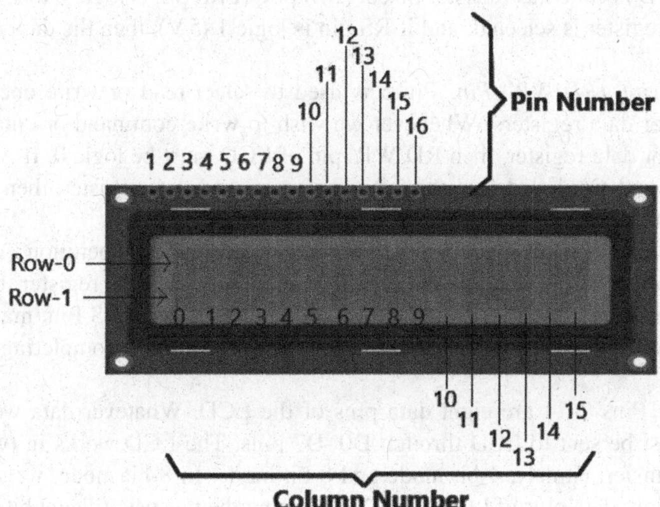

FIGURE 3.13 LCD module.

TABLE 3.4
Pin Description of LCD

Pin Number	Name of Pin	Description
1	Gnd	Ground (0 V)
2	V_{CC}	Supply voltage (5 V)
3	V_{EE}	Contrast adjustment
4	Register Select (RS)	If RS = 0, then select command register If RS = 1, then select data register
5	RD/WR′	If RD/WR′ = 0, then write to the register If RD/WR′ = 1, then read from the register
6	Enable (EN)	A high-to-low pulse on Enable pin is required for "write" operation.
7	D0	Eight data pins
8	D1	
9	D2	
10	D3	
11	D4	
12	D5	
13	D6	
14	D7	
15	Backlight (5 V)	+Vcc
16	Backlight (GND)	GND

Once proper command word is loaded in the command register, and then only the data which we want to display should be loaded in the data register.

To select the command register or data register, the Pin 4 of the LCD is used. Pin 4 of the LCD is called as Resister Select (RS) pin. If RS pin is logic 0 (0 V), then the command register is selected, and if RS pin is logic 1 (5 V), then the data register is selected.

Read/Write′ (RD/WR′) Pin: Pin 5 is used to select read or write operation on command or data registers. Whenever we wish to write command or data word in command or data register, then RD/WR′ pin of LCD must be logic 0. If we wish to read command word or data word from command and data register, then RD/WR′ pin of LCD must be logic 1.

Enable (En): Pin 6 is used to enable the read or write operation. Whenever we want to write command or data word in command or data register, then after selecting proper register by using appropriate voltage level at RS Pin, making RD/WR′ = logic 0 a high-to-low pulse on Enable pin is required for completing the write operation.

D0–D7: Pins 7–14 are eight data pins of the LCD. Whatever data we want to display must be sent to LCD through D0–D7 pins. The LCD works in two modes for data transfer, namely, 4-bit mode and 8-bit mode. In 8-bit mode, we send 8-bit data over eight data pins of LCD with D7 receiving the most significant bit (MSB) of data and D0 receiving the least significant bit (LSB) of data. In the 8-bit mode of data transfer, we need eight pins of LCD from D0 to D7.

In 4-bit mode, we send 8-bit data in the form of two nibbles (4 bit). First, we have to send upper nibble and then lower nibble of data. In the 4-bit mode of data transfer, we need four pins of LCD from (D4 to D7).

Backlight Illumination: Pins 15 and 16 are used to illuminate the backlight of the LCD. Pin 15 is connected to 5 V, and Pin 16 is connected to the ground.

* *Refer to Section 4.6 of Chapter 4 for programming and interfacing of the LCD with Arduino UNO.*

3.5 POTENTIOMETER

A potentiometer is a three-terminal variable resistor device whose resistance can be changed manually. It is also called pot and available in the market in a variety of shapes. A potentiometer is shown in Figure 3.14. The symbol of a potentiometer is shown in Figures 3.15 (a) and 3.15 (b), and users can use any one symbol. The mark "B10K" on the potentiometer, as shown in Figure 3.14, indicates that the potentiometer's maximum resistance is 10 KΩ. The alphabet "B" on the mark "B10K" indicates that the potentiometer's resistance varies linearly as Terminal 2 moves away from Terminal 3 to Terminal 1 (or from Terminal 1 to Terminal 3).

FIGURE 3.14 The image of a 10 KΩ potentiometer.

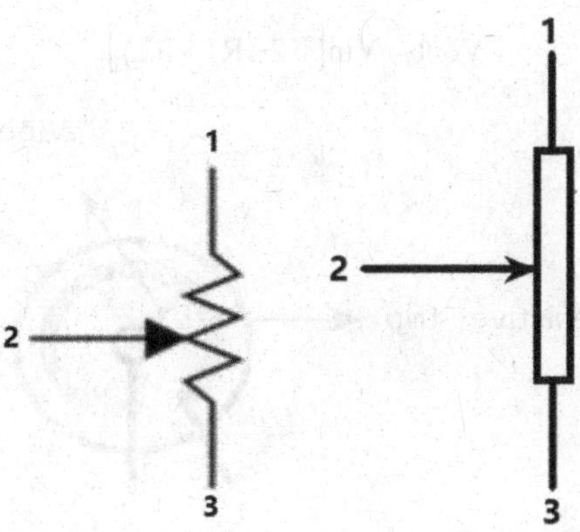

FIGURE 3.15 (a) Potentiometer Symbol 1. (b) Potentiometer Symbol 2.

3.5.1 Internal Schematic of Potentiometer

A potentiometer is a three-terminal variable resistor. The three terminals are as follows:

Terminal 1: It is named Vin, and the positive terminal of a voltage source is connected to this terminal.
Terminal 3: It is named Gnd, and the ground terminal of a voltage source is connected to this terminal.
Terminal 2: It is named wiper, and this terminal is movable.

Terminal 1 and Terminal 3 are interchangeable; i.e., we can connect the positive terminal of a voltage source to Terminal 1, and the ground terminal of a voltage source to Terminal 3 or the reverse is also possible.

The internal schematic of a potentiometer is shown in Figure 3.16. Internally, there is a resistive strip whose resistance varies as we move away from one end to another. It can be observed that the Terminal 2 (wiper) is more close to Terminal 3 in Figure 3.17 (a). If resistance is measured between the Terminal 2 (wiper) and Terminal 3, it will be less than the resistance measured between the Terminal 2 (wiper) and Terminal 1. Figure 3.17 (b) shows that the Terminal 2 (wiper) is closer to Terminal 1. If resistance is measured between the Terminal 2 (wiper) and Terminal 3, it will be more than the resistance measured between the Terminal 2 (wiper) and Terminal 1.

3.5.2 Working of Potentiometer

Regarding Figure 3.18, let resistance between Terminal 2 (wiper) and Terminal 1 be represented by R1, and the resistance between Terminal 2 (wiper) and Terminal 3 be represented by R2.

Regarding the circuit shown in Figure 3.19, the output voltage measured between the Terminal 2 (wiper) and Terminal 3 (Gnd) is calculated by using the potential divider rule and is given by (3.1):

$$\text{Vout} = \text{Vin}\left[\text{R2}/(\text{R1}+\text{R2})\right] \tag{3.1}$$

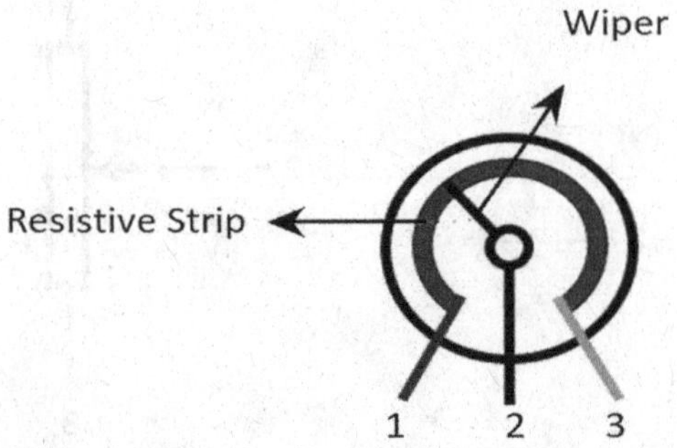

FIGURE 3.16 Internal schematic of potentiometer.

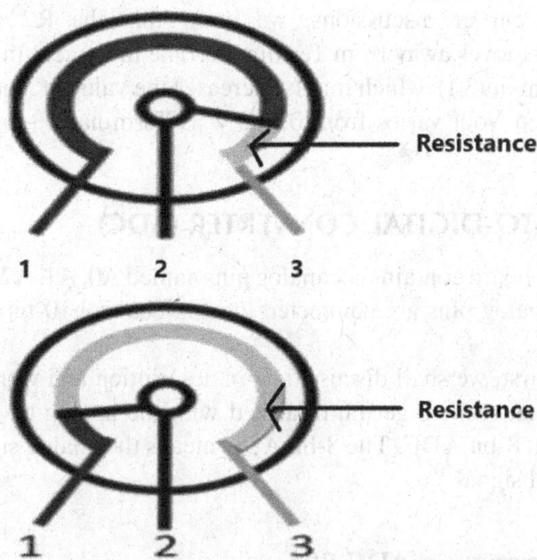

FIGURE 3.17 (a) Potentiometer when the resistance between Terminal 2 and Terminal 3 is less than the resistance between Terminal 2 and Terminal 1. (b) Potentiometer when the resistance between Terminal 2 and Terminal 3 is more than the resistance between Terminal 2 and Terminal 1.

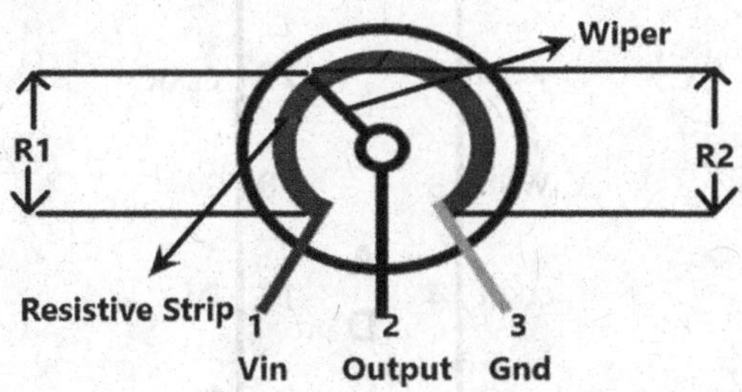

FIGURE 3.18 Internal schematic of potentiometer with resistances R1 and R2.

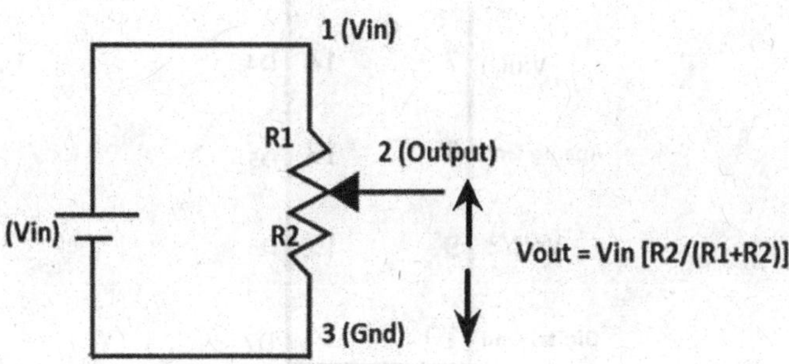

FIGURE 3.19 Circuit diagram of a potentiometer.

Regarding our earlier discussions, we know that the R2 will increase as Terminal 2 (wiper) moves away from Terminal 3. The increment in R2 will increase [R2/(R1 + R2)] term of (3.1), which finally increases the value of Vout. Let us assume that Vin is 5 V, then Vout varies from 0 to 5 V as Terminal 2 (wiper) moves away from Terminal 3.

3.6 ANALOG-TO-DIGITAL CONVERTER (ADC)

The Arduino UNO board contains six analog pins named A0, A1, A2, A3, A4, and A5. Internally, these analog pins are connected to a six-channel 10-bit analog-to-digital converter.

In this section first, we shall discuss the pin description and working principle of ADC 804 IC so that readers get familiarized with the analog-to-digital converter. The ADC 804 is an 8-bit ADC. The 8-bit ADC means the analog signal is converted into an 8-bit digital signal.

3.6.1 Pin Description of ADC 804

The pin diagram of ADC 804 is shown in Figure 3.20.

The function of various pins of ADC 804 is discussed as follows:

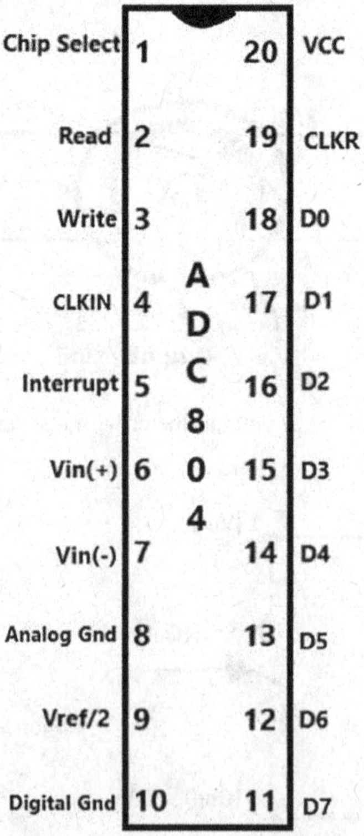

ADC804

Label	Pin	Pin	Label
Chip Select	1	20	VCC
Read	2	19	CLKR
Write	3	18	D0
CLKIN	4	17	D1
Interrupt	5	16	D2
Vin(+)	6	15	D3
Vin(-)	7	14	D4
Analog Gnd	8	13	D5
Vref/2	9	12	D6
Digital Gnd	10	11	D7

FIGURE 3.20 Pin diagram of ADC 804.

Vcc: A DC voltage of 5 V is required to be connected at Vcc pin for operation.
Vin(+): The analog signal which has to be converted into digital form is connected at this pin.
Vin(−): This pin is connected to the ground.

The signal which has to be converted into digital form is given by (3.2):

$$\text{Vin} = \text{Vin}(+) - \text{Vin}(-) \tag{3.2}$$

Since Vin(−) is connected to grbund (0 V), (3.2) becomes:

$$\text{Vin} = \text{Vin}(+) \tag{3.3}$$

Thus if we connected Vin(−) to the ground, then the analog signal connected to Vin(+) pin will be converted into a digital signal.

Vref/2: This pin is used for allowing a specific voltage range at Vin(+) pin for the conversion of an analog signal into a digital signal. If Vref/2 is open, we can apply a voltage in the range of 0–5 V to Vin(+) to convert it into a digital signal. If Vref/2 is connected to 2 V, then we can apply a voltage in the range of 0–4 V to Vin(+) for converting it into a digital signal.

Suppose we are using a sensor in an application and the sensor is expected to generate the analog signal in the range 0–3 V. If we wish to convert the sensor's analog signal into digital data, then we shall connect the analog signal at Vin(+) pin of ADC 804. In this case, the expected analog signal generated by the sensor is in the range 0–3 V; therefore, we should connect the Vref/2 pin of ADC to 1.5 V.

CLKIN: When the analog signal is connected to Vin input for conversion into a digital signal, the conversion will not instantly occur. It needs some time to convert an analog signal into digital. A clock signal is required for providing the timing requirement, and this clock signal is to be connected at CLKIN pin of ADC 804. If we connect a clock of 606 kHz, then the analog-to-digital conversion will be completed in 110 μs. For ADC 804, 110 μs is the fastest conversion time, which can be achieved if a clock of 606 kHz is connected to the CLKIN pin. If the clock connected to the CLKIN pin is less than 606 kHz, then converting an analog signal into a digital signal will take more than 110 μs.

CLKR: If the clock signal is not readily available for the CLKIN pin of ADC 804, then the ADC's internal oscillator can be initiated to generate a clock signal by connecting a resistor and a capacitor to CLKIN and CLKR pins as shown in Figure 3.21. The frequency of clock signal generated will be given by (3.4):

$$F = 1/1.1RC \tag{3.4}$$

If we select R= 10 kHz for F= 606 kHz, then C's value can be calculated from (3.4) as 150 pF. Thus to achieve a conversion time of 110 μs, we need 606 kHz clock frequency, and this can be done by selecting R = 10 KHz and C = 150 pF.

Analog Ground: This pin is connected to the analog circuit's ground signal, which is interfaced with ADC.

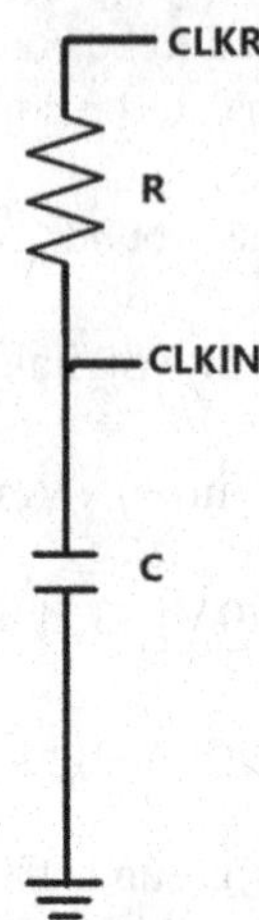

FIGURE 3.21 A resistor and capacitor connection to CLKIN and CLKR pins to initiate internal oscillator of ADC 804.

Digital Ground: This pin is connected to the digital circuit's ground signal, which is interfaced with ADC.

D7–D0: These are eight digital data pins of ADC 804. After converting the analog signal, the digital signal is available on these eight digital data pins.

Chip Select (CS′): If we wish to use ADC 804, it must first be selected. If CS′ = logic 0, then ADC 804 will be selected.

Write: This pin is also called as "Start of Conversion" (SOC) pin. It is an input pin of ADC. A LOW to HIGH pulse at this pin initiates the start of the conversion of an analog signal which is connected to the Vin(+) pin of ADC into a digital signal.

Interrupt: This pin is also called as "End of Conversion" (EOC) pin. It is an output pin of ADC, and by default, it is at a high (5 V/logic 1) value. When ADC completes an analog signal's conversion into a digital signal, this pin goes low (0 V/logic 0) for a small duration and again comes back to the high level. If we observe interrupt pin continuously, we can determine when the analog-to-digital conversion of data is completed.

Read: Once the analog signal is converted into a digital signal, it is in ADC's latch. It is still not available on D7–D0 digital data lines. The read signal must go from high to low to make digital data available on D7–D0 digital data lines. Once the digital data is available on D7–D0 pins, they can be used in the program for processing.

3.6.2 Analog to the Digital Data Conversion Process in ADC 804

The function of write, interrupt, and read signals during the conversion of analog-to-digital data in ADC 804 is shown in Figure 3.22.

After receiving the analog data for conversion into digital data, a low-to-high signal on the "Write" pin of ADC will initiate the conversion process. If the ADC is operated by 606 kHz clock signal, then the conversion process will complete in 110 μs, and once conversion completes signal on the "Interrupt", pin goes from high

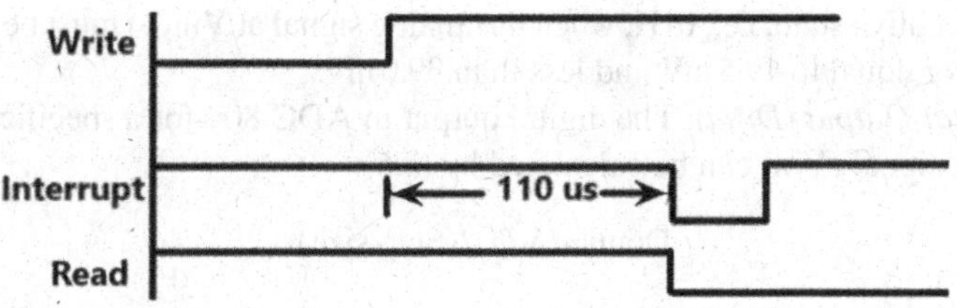

FIGURE 3.22 Write, Interrupt, and Read signals of ADC 804.

to low level. A high-to-low signal on the "Read" pin will make converted digital data available on D7–D0 pins for further processing.

3.6.3 Important Terminology of ADC

Important terms related to analog-to-digital converter are as follows:

i. *Step Size*: The minimum amount of analog signal required to send the digital output of an ADC to the next consecutive state is called step size. Let us take an example of ADC 804. The step size is represented by (3.5):

$$\text{Step size} = (\text{Maximum allowed analog input})/(\text{Number of steps}) \quad (3.5)$$

Here, Number of steps = 2^n, where n = Number of digital output lines.

Since the number of digital output lines in ADC 804 is eight, the number of steps in (3.5) is 256, i.e., from 0 to 255. Let us assume the maximum allowed analog input in an application is 5 V; therefore, the step size will be calculated using (3.5), and its value is 19.5 mV. The relation between step size and digital output of ADC 804 is shown in Table 3.5.

If the analog signal at Vin(+) is 0 V, then the digital output is 00H (H indicates hexadecimal number system). The digital output goes to the next

TABLE 3.5
The Relation between Step Size and Digital Output of ADC 804

No. of Step	Vin	Digital Output D7	D6	D5	D4	D3	D2	D1	D0	Hexadecimal Value Equivalent to Digital Output
0	0V	0	0	0	0	0	0	0	0	00H
1	19.5 mV	0	0	0	0	0	0	0	1	01H
2	39.0 mV	0	0	0	0	0	0	1	0	02H
3	58.5 mV	0	0	0	0	0	0	1	1	03H
–										
–										
–										
255	5V	1	1	1	1	1	1	1	1	FFH

consecutive state, i.e., 01H, when the analog signal at Vin(+) must be greater than or equal to 19.5 mV and less than 39.0 mV.

ii. *Digital Output (Dout)*: The digital output in ADC 804 for a specific analog input signal (Vin) can be calculated by (3.6):

$$\text{Dout} = (\text{Vin})/(\text{Step size}) \tag{3.6}$$

Example 3.3

For an ADC 804 if the Vref/2 pin is open, then find out the digital output signal when the analog input signal is 160 mV.

Solution

Since in the ADC 804 Vref/2 pin is open, the maximum allowed analog input is 5 V; therefore, the step size will be calculated using (3.5), and its value is 19.5 mV.

It is given that Vin = 160 mV, therefore, from (3.6):

$$\text{Dout} = (\text{Vin})/(\text{Step size})$$

or, $\text{Dout} = (160\,\text{mV})/(19.5\,\text{mV})$

or, Dout = 8.3

or, Dout = 8 (after rounding off)

or, Dout = 00001000 (representing 8 in eight bit binary form)

3.6.4 Analog Inputs in Arduino UNO Board

The Arduino UNO board contains six analog pins named A0, A1, A2, A3, A4, and A5. Internally, these analog pins are connected to a six-channel 10-bit analog-to-digital converter. The allowable analog input voltage range at each analog input pin is 0–5 V. Since each analog input pin is connected to a 10-bit analog-to-digital converter, 0–5 V is divided into 1,024 steps. So there are 1,024 steps ranging from 0 to 1,023 steps. The six analog pins are shown in Label 10 of Figure 1.1 (Chapter 1).

The minimum amount of analog signal required to send ADC's digital output to the next consecutive state is called step size. Let us take an example of ADC 804. The step size is represented by (3.5). The step size for the ADC of the Arduino UNO board will be 4.9 mV.

* *Refer to Section 4.7 of Chapter 4 for programming and interfacing of the potentiometer with Arduino UNO.*

3.7 PULSE WIDTH MODULATION (PWM)

The pulse width modulation (PWM) is a technique by which we can indirectly encode the digital value into an equivalent analog value.

Using PWM, we can generate some rectangular digital waveform, and there will be a particular period T. When the rectangular digital waveform is high, we say it is on.

When it is low, we say it is off. The matter of interest is that for how long waveform is on and for how long it is off. The output signal will alternate between on and off durations with a specific period; of course, this on-time and off-time will also be fixed for a particular case. The on–off behavior changes the average power of the signal.

3.7.1 Working of PWM Concept

In PWM, the period of the waveform is kept constant. Once we specify this period T, this will be kept constant. Depending on the parameter, we want to encode something we are trying to vary. The pulse width or the on-time let us call it Ton; this Ton is something we are varying in proportion to the parameter we want to encode.

In this respect, we define something called the duty cycle of this rectangular waveform. It is the proportion of time the pulse is on expressed as a percentage. So the actual duty cycle is defined as the total time for which our pulse is high divided by the total period or the pulse period multiplied by 100. A rectangular digital waveform is shown in Figure 3.23. The duty cycle is given by (3.7):

$$\begin{aligned}\text{Duty cycle} &= \left[(\text{Pulse on time})/(\text{Pulse time period})\right]\times 100\% \\ &= (\text{Ton}/\text{T})\times 100\% \end{aligned} \tag{3.7}$$

The Fourier transform of this waveform can calculate the DC component, but we shall not go in that mathematical solution.

Let us assume this voltage is varying from 0 to 5 V. This dotted line in Figure 3.23 shows the average value. This average value will very much depend on how much time the pulse is on. More the on period, this dotted line will be moving up, but if the pulses are very narrow, this dotted line will move down.

So, by adjusting Ton, the average value will indirectly be proportional to the on period or the duty cycle. If the period is constant, the duty cycle is proportional to Ton, which is proportional to the average value.

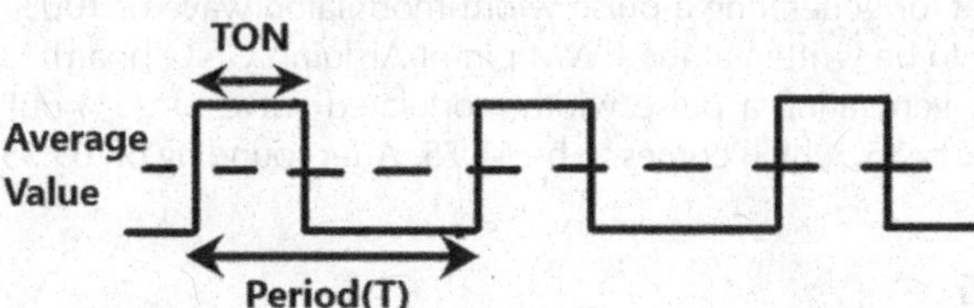

FIGURE 3.23 A rectangular digital waveform

3.7.2 Applications of PWM

Some typical applications of PWM are given below:

i. Suppose we are trying to control the speed of a DC motor. There is often a control signal with the help of which we can adjust the power supply we are applying to the motor.

If the PWM waveform we are applying directly, then the average value will be the equivalent power supply. So by adjusting the duty cycle, we are effectively adjusting the power supply, thereby adjusting the motor's speed.

ii. By adjusting the duty cycle of a wave, we adjust the power of the microwave oven.

iii. In an automatic heater control system, we turn on and off the heater; the time we are turning on the heater will be proportional to the PWM waveform's duty cycle.

3.7.3 PWM Pins in Arduino UNO Board

The Arduino UNO board contains six pins, namely, 3, 5, 6, 9, 10, and 11, with PWM capabilities. The six PWM pins are labeled as "~".

The *`analogWrite(pin, value)`* function is used to write an analog value to PWM pins. The *`analogWrite`* function has two arguments. The first argument is *`pin`*, and the second argument is *`value`*. The argument *`pin`* is used to specify the PWM pin from which we wish to generate an analog output. The argument *`value`* is used to control the duty cycle of the pulse generated from the specified PWM pin. The argument *`value`* can have a value from 0 to 255.

We can vary the duty cycle of a pulse from 0% to 100% by varying the *`value`* from 0 to 255 to generate an analog voltage from 0 to 5 V.

Example 3.4

Calculate the *`value`* required for generating a pulse width-modulated wave of 25% duty cycle at PWM Pin 9 of the Arduino UNO board in the given function:

```
analogWrite(9,value);
```

Solution

We know that for generating a pulse width-modulated wave of 100% duty cycle, we need 255 to be written at the PWM pin of Arduino UNO board.

Therefore, generating a pulse width-modulated wave of 25% duty cycle, we need (255/100)×25, which comes to be 63.75. After rounding off 63.75, we get 64.

Example 3.5

How much analog voltage will be generated at PWM Pin 9 of the Arduino UNO board of Example 3.4 if a 25% duty cycle pulse is generated at Pin 9 of Arduino board.

Solution

We know that for a wave of 100% duty cycle, 5 V is generated. Therefore, 25% duty cycle wave will generate (5/100)×25 V, which comes to be 1.25 V.

** Refer to Section 4.8 of Chapter 4 for Arduino UNO programming using PWM.*

3.8 TEMPERATURE SENSOR LM35

The LM35 is a precision temperature sensor IC. The LM35 temperature sensor IC is shown in Figure 3.24. It generates an output voltage that is linearly proportional to the temperature in degree centigrade. The LM35 is a three-pin IC, and the functions of these pins are as follows:

i. *Pin 1*: It is to be connected to a positive power supply. In LM35, this power supply voltage can vary between 4 and 20 V.
ii. *Pin 2*: An analog voltage will be generated from this pin that will be proportional to the external ambient temperature. The generated voltage is 10 mV/°C and linearly proportional to the temperature in degree centigrade or Celsius.
iii. *Pin 3*: It is to be connected to the ground.

LM35 temperature sensor does not need any external calibration; it has all calibration circuitry built inside, and we do not need any external calibration. LM35 has an accuracy of ±0.25°C, which is often sufficient in most applications. It consumes very low power, 60 μA current during operation. It is also a very low-power device, and it can operate over a wide range of temperatures from −55°C to +155°C.

Some other temperature sensor ICs are available in the LM35 series whose output voltage is proportional to the temperature in degree Fahrenheit. LM34 generates an output voltage of 10 mV/°F.

* *Refer to Section 4.9 of Chapter 4 for programming and interfacing of the LM35 with Arduino UNO.*

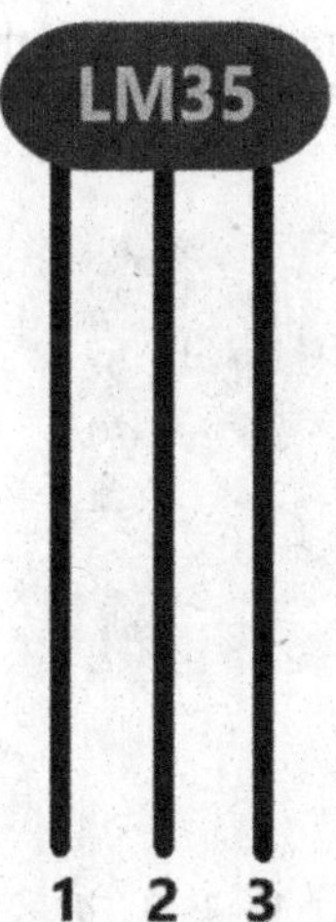

FIGURE 3.24 LM35 temperature sensor.

3.9 HUMIDITY AND TEMPERATURE SENSOR DHT11

The DHT 11 is a digital humidity and temperature sensor. This sensor is easy to use and easily interfaced with the Arduino UNO board or any other microcontroller. It can measure humidity and temperature instantly. To measure the humidity, it uses a capacitive humidity sensor, and to measure the temperature, DHT11 uses a thermistor. The analog signal will be generated in response to the measured humidity and temperature value, but inside DHT11, the analog signal is converted into a digital signal and sent out from the data pin of DHT11.

3.9.1 Pin Description of DHT11

The DHT11 is a four-pin IC, as shown in Figure 3.25, and the functions of these pins are as follows:

i. *Pin 1*: It is to be connected to a positive power supply. In DHT11, this power supply voltage can vary between 3.5 and 5.5 V.
ii. *Pin 2*: This is the data pin for both measured humidity and temperature values. The equivalent digital signal of humidity and temperature will be sent out serially from this pin.
iii. *Pin 3*: It is no connection pin. This pin must be kept open for the regular operation of DHT11.
iv. *Pin 4*: It is to be connected to the ground.

3.9.2 General Features of DHT11

The humidity range and the temperature range of the DHT11 sensor are from 20% to 80% with ±5% accuracy and 0°C–50°C with ±2°C accuracy. The sampling rate of this

FIGURE 3.25 DHT 11 temperature and humidity sensor.

sensor is 1 Hz; i.e., it gives one reading every second. DHT11 is small in size with an operating voltage from 3.5 to 5.5 V. The maximum current used while requesting data is 2.5 mA. The sensor data output Pin 2 is open collector; therefore, a pull-up resistor of (5–10) KΩ is required to be connected between Vcc and data Pin 2. Since Pin is always connected to 5 V, we can connect the pull-up resistor between Pins 1 and 2.

3.9.3 Working Principle of DHT11

DHT11 sensor consists of a capacitive humidity-sensing element and a thermistor for sensing temperature. The humidity is sensed by using a capacitive sensor. A capacitive sensor converts a nonelectrical quantity (e.g., force, pressure, humidity) into an electrical quantity (e.g., voltage or current) through capacitance change. The principle of operation of a capacitive sensor is based on the equation of capacitance of a parallel plate capacitor (3.8).[2]

$$C = \varepsilon_r \varepsilon_o A/d \tag{3.8}$$

Here, C = Capacitance of a capacitor
ε_r = Relative permittivity
ε_o = Permittivity of free space = 8.85×10^{-12} Farad/meter
A = Area of the overlapping parallel plate (m^2)
D = Distance between two parallel plates (m)

The dielectric material is selected such that it can hold the moisture. When the humid air enters the humidity sensor, the humid air will be held by the dielectric material, and the capacitance of the capacitor will change. Due to the capacitance change, the capacitive reactance (Xc) of the capacitor will change. The capacitive reactance (Xc) of the capacitor is given by (3.9):

$$Xc = 1/2\pi fC \tag{3.9}$$

Here, Xc = Capacitive reactance (Ω)
f = Frequency of signal (Hz)
C = Capacitance of capacitor (Farad)

The change in the capacitive reactance will be recorded with the change in humidity levels. The voltage-level change recognizes the change in humidity level, and this voltage is finally converted into digital form and available at data Pin 2 of DHT11.

The temperature sensor is made up of a material that shows a negative temperature coefficient property. If the material has a negative temperature coefficient property, its resistance decreases with temperature. A temperature sensor made up of a material that shows a negative temperature coefficient is called a thermistor. The semiconductor materials show the negative temperature coefficient property. The change in the resistance will be recorded with the change in temperature levels. The voltage-level change recognizes the change in temperature level, and this voltage is finally converted into the digital form and available at data Pin 2 of DHT11.

3.9.4 Timing Diagram of DHT11

The timing diagram when the DHT11 humidity and temperature sensor is accessed for measuring the humidity and temperature is shown in Figure 3.26. The various states of the timing diagram are as follows:

State 1: In the idle state, the signal is pulled HIGH up by the resistor to Vcc.
State 2: The processor pulls down the signal low (0 V/logic 0) for at least 18 ms to inform the sensor that it shall send the data.
State 3: The processor releases the pin, and the signal will be pulled up again.
State 4: The sensor responds by pulling down the signal low for 80 µs (Start LO) followed by 80 µs high (Start HI).
State 5: The sensor sends 40 data bits, with a binary 1 indicated by high for 70 µs and a binary 0 indicated by high for 26 µs.
State 6: In the end, the signal is released and pulled high again.

3.9.5 Data Format of DHT11

The 40 bits of data of DHT11 are organized, as shown in Figure 3.27. The 16 bits are reserved for representing humidity. The first 8 bits of humidity data represent the integral part and the second 8 bits the fractional part. For the DHT11, the fractional bits are always zero. The 16 bits are reserved for representing temperature. Like

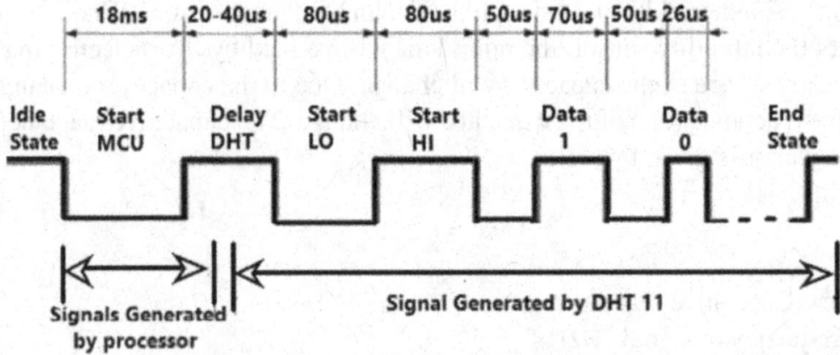

FIGURE 3.26 The timing diagram when DHT11 humidity and temperature sensor is accessed for measuring the humidity and temperature.

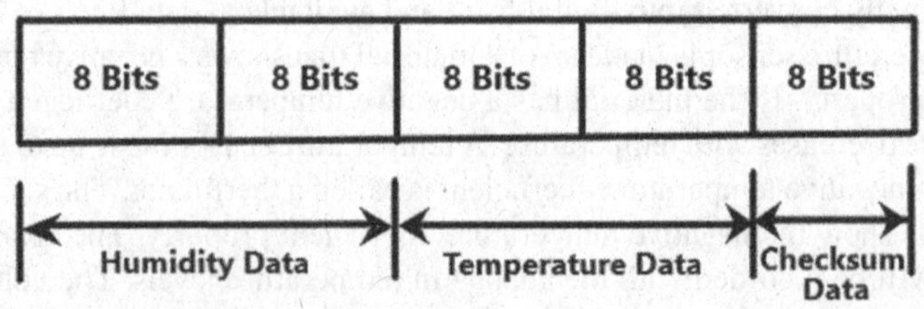

FIGURE 3.27 The 40 bits of data of DHT11.

humidity data, the 8 bits represent an integral part of temperature, and the second 8 bits represent the fractional data and the fractional data is always zero for the DHT11. The 8 bits are checksum bits.

* *Refer to Section 4.10 of Chapter 4 for programming and interfacing of the humidity and temperature sensor DHT11 with Arduino UNO.*

3.10 MOTOR DRIVER L293D

A motor driver IC is an integrated chip used as an interface between a motor and microprocessors/microcontrollers. The current capability of microprocessors/microcontrollers is small; however, motor requires more current for its operation; therefore, the motor driver fulfils the current requirement by enhancing it.

3.10.1 Pin Description of L293D

The L293D, a 16-pin motor driver IC, belongs to the family of L293. The L293D IC is shown in Figure 3.28, and the pin diagram is shown in Figure 3.29. The L293D IC has to deal with the large current; therefore, four ground pins are provided: Pins 4, 5, 12, and 13. The description of pins of L293D IC is as follows:

Pin 1: ENABLE 1–2; this pin is used to activate INPUT 1 (Pin 2) and INPUT 2 (Pin 7).

If ENABLE 1–2 = 0, then INPUT 1 and INPUT 2 are disabled, and if ENABLE 1–2 = 1, INPUT 1 and INPUT 2 are enabled.

Pin 2: INPUT 1, when this pin is HIGH, the current will flow through OUTPUT 1 provided INPUT 1 is already enabled by making ENABLE 1–2 = 1.

Pin 3: OUTPUT 1, this pin is to be connected to one terminal of the motor.

Pins 4, 5: GND, these pins are ground pins and connected to the ground.

Pin 6: OUTPUT 2, this pin is to be connected to one terminal of the motor.

Pin 7: INPUT 2, when this pin is HIGH, the current will flow through OUTPUT 2 provided Input 1 is enabled by making ENABLE 1–2 = 1.

Pin 8: Vcc2, the voltage connected to this pin will be supplied to the motor. For example, if we are driving a 12 V DC motor, this pin must be connected to 12 V.

Pin 9: ENABLE 3–4; this pin is used to activate INPUT 3 (Pin 10) and INPUT 4 (Pin 15).

Pin 10: INPUT 3, when this pin is HIGH, the current will flow through OUTPUT 3 provided INPUT 3 is already enabled by making ENABLE 3–4 = 1.

Pin 11: OUTPUT 3, this pin is to be connected to one terminal of the motor.

Pins 12, 13: GND, these pins are ground pins and connected to the ground.

Pin 14: OUTPUT 4, this pin is to be connected to one terminal of the motor.

Pin 15: INPUT 4, when this pin is HIGH, the current will flow through OUTPUT 4 provided INPUT 4 is enabled by making ENABLE 3–4 = 1.

Pin 16: Vcc1, this pin is the power source to the IC and used for IC's internal operation. So this pin should be connected to 5 V.

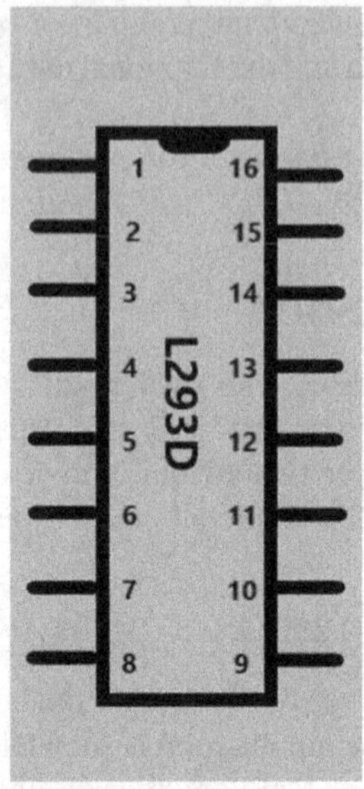

FIGURE 3.28 The L293D IC.

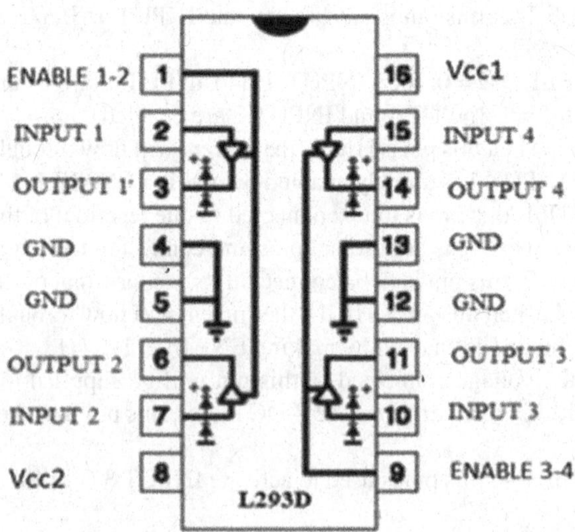

FIGURE 3.29 Pin diagram of L293D motor driver IC.

3.10.2 Working of L293D

Let us assume that we have a 5 V DC motor with two terminals Terminal 1 and Terminal 2. It requires a maximum of 5 V to rotate motor at its maximum speed if the supplied voltage is less than 5 V motor rotates at the lower speed. A simple 5 V motor and its symbol are shown in Figures 3.30 and 3.31, respectively. The rotation status of motor M1 to the voltages on Terminal 1 and Terminal 2 is shown in Table 3.6.

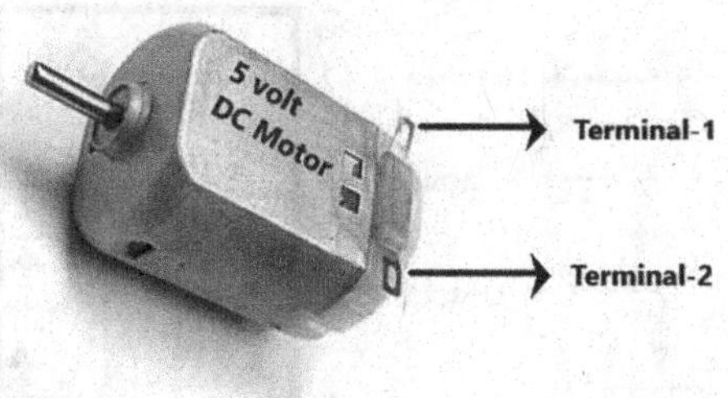

FIGURE 3.30 A 5 V DC motor.

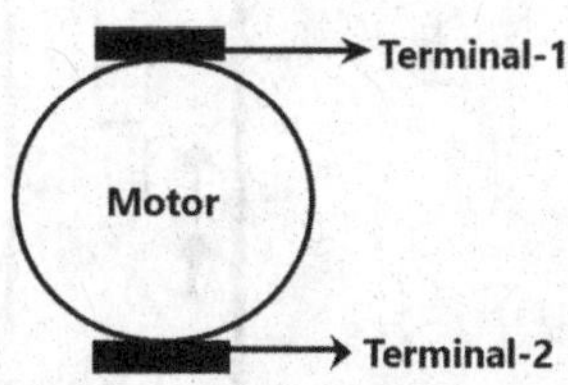

FIGURE 3.31 Symbol of motor.

TABLE 3.6
Working of DC Motor

Terminal 1	Terminal 2	Rotation
0	0	Stop
0	1	Anti-clockwise
1	0	Clockwise
1	1	Stop

The working principle of L293D is explained under the following cases:

Case 1(a): In Case 1(a), the Terminal 1 and Terminal 2 of the motor M1 are connected to Output 1 and ground as shown in Figure 3.32 and various conditions of rotation of the motor in Case 1(a) if Enable 1–2 is at Logic 1 are shown in Table 3.7. If Terminal 1 is connected to Logic 1, then the motor starts rotating in the clockwise direction, and if it is connected to Logic 0, then the motor stops. There is no way to rotate the motor in the anticlockwise direction because the Terminal 2 of the motor is connected permanently to the ground.

Case 1(b): In Case 1(b), the Terminal 1 and Terminal 2 of the motor M1 are connected to the ground and Output 1 as shown in Figure 3.33, and various conditions of rotation of the motor in Case 1(b) if Enable 1–2 is at Logic 1 are shown in Table 3.8. If Terminal 2 is connected to Logic 1, then the motor starts rotating in the anticlockwise direction, and if it is connected to Logic

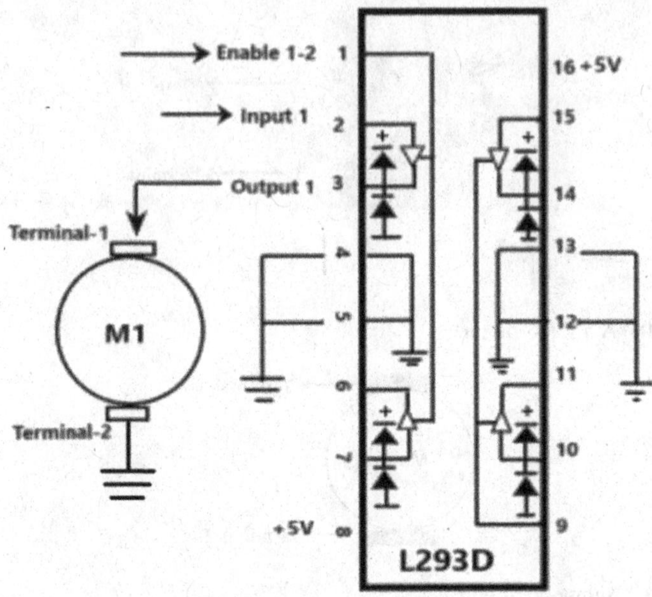

FIGURE 3.32 Motor connection with L293D motor driver IC for Case 1 (a).

TABLE 3.7
Working of L293D for Case 1(a)

L293D		Motor (M1)		
Input 1	**Output 1**	**Terminal 1**	**Terminal 2**	**Rotation**
0	0	0	0	Stop
1	1	1	0	Clockwise

0, then the motor stops. There is no way to rotate the motor in a clockwise direction because the Terminal 1 of the motor is connected permanently to the ground.

Conclusion

i. In Case 1, we can rotate a motor either in the clockwise direction or in an anticlockwise direction. Still, it is not possible to rotate a motor in both clockwise and anticlockwise directions.
ii. We can connect a maximum of four motors, but we can only rotate these motors in one direction either in the clockwise direction or in the anticlockwise direction. One method to interface four motors with L293D is shown in Figure 3.34.

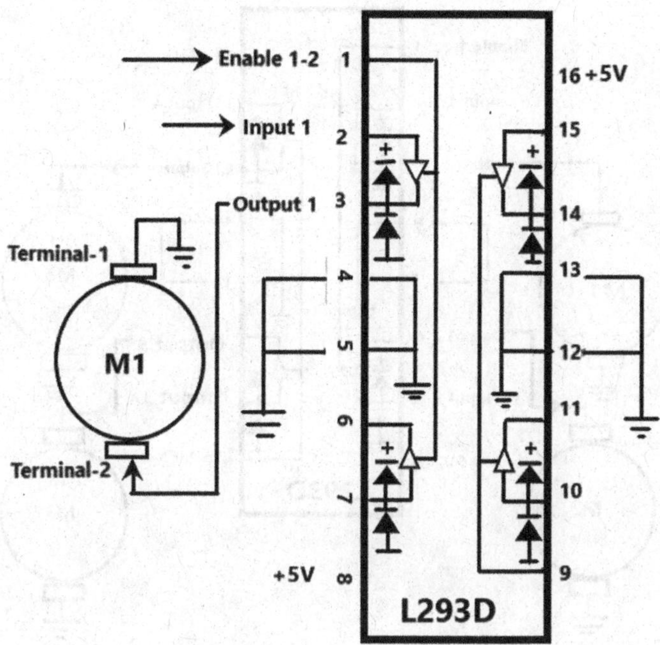

FIGURE 3.33 Motor connection with L293D motor driver IC for Case 1 (b).

TABLE 3.8
Working of L293D for Case 1(b)

L293D		Motor (M1)		
Input 1	**Output 1**	**Terminal 1**	**Terminal 2**	**Rotation**
0	0	0	0	Stop
1	1	0	1	Anti-clockwise

Case 2: In Case 2, the Terminal 1 and Terminal 2 of the motor M1 are connected to the Output 1 and Output 2, respectively, as shown in Figure 3.35 and various conditions of rotation of the motor in Case 2 if Enable 1–2 is at Logic 1 are shown in Table 3.9. If Terminal 1 and Terminal 2 are connected to Logic 0 and Logic 1, then the motor starts rotating in an anticlockwise direction. If Terminal 1 and Terminal 2 are connected to Logic 1 and Logic 0, respectively, the motor starts rotating in the clockwise direction. If both the terminals are either at Logic 0 or at Logic 1, then the motor stops.

Conclusion

i. In Case 2, we can rotate a motor in either clockwise direction or anticlockwise direction.

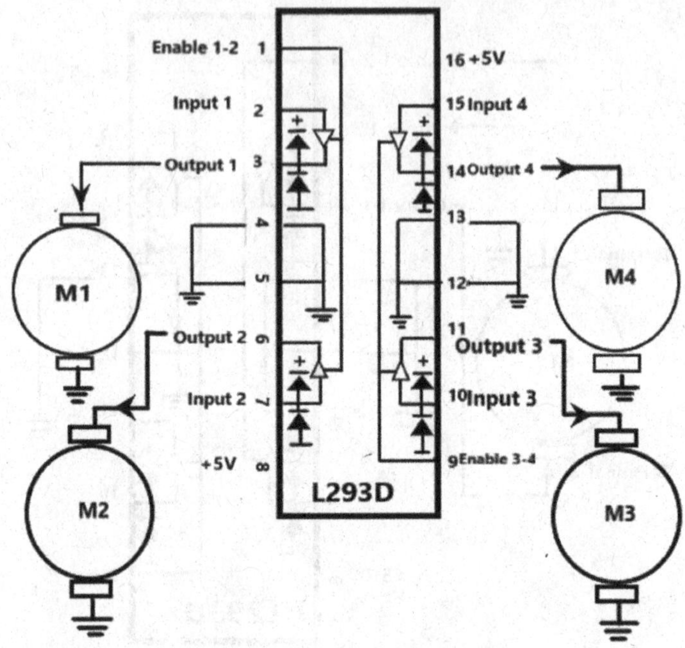

FIGURE 3.34 The interfacing of four motors with L293D motor driver.

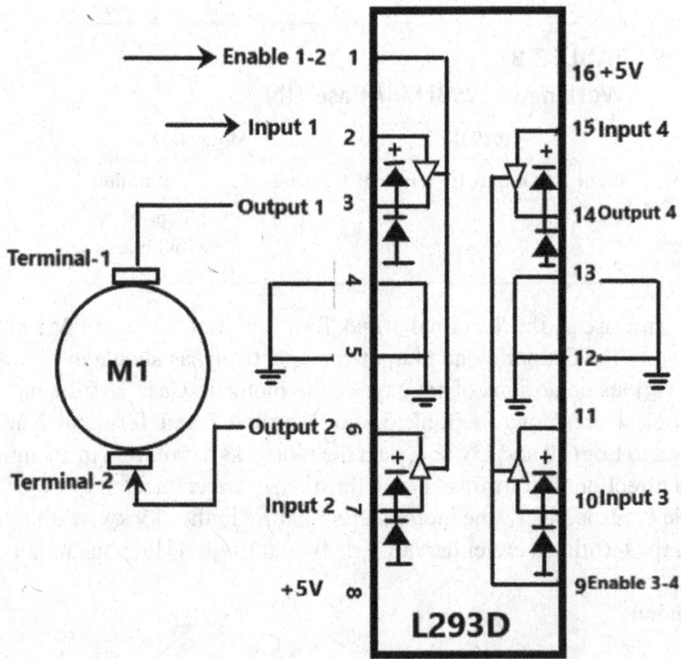

FIGURE 3.35 The interfacing of one DC motor with L293D motor driver IC in Case 2.

TABLE 3.9
Working of L293D for Case 2

L293D				Motor (M1)		
Input 1	**Output 1**	**Input 2**	**Output 2**	**Terminal 1**	**Terminal 2**	**Rotation**
0	0	0	0	0	0	Stop
0	0	1	1	0	1	Anti-clockwise
1	1	0	0	1	0	Clockwise
1	1	1	1	1	1	Stop

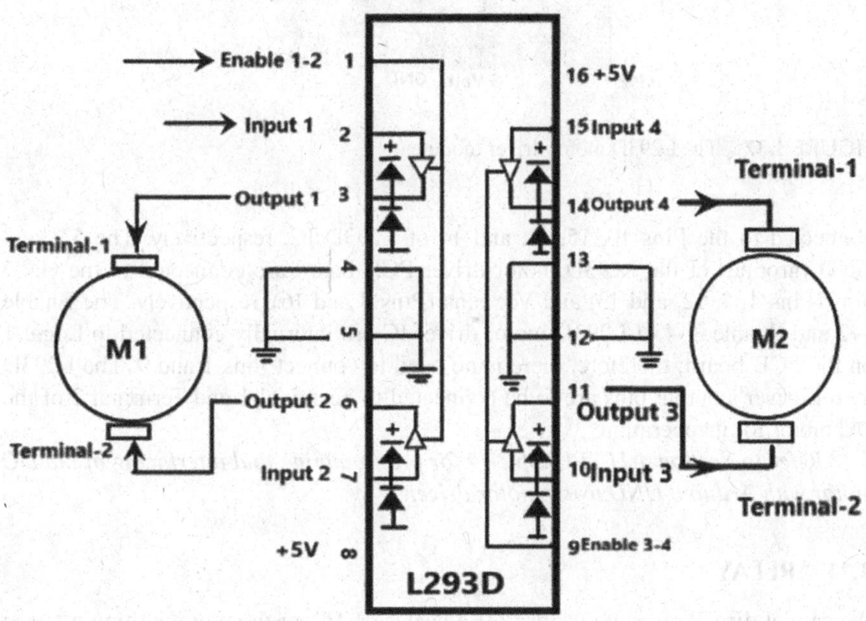

FIGURE 3.36 The interfacing of two DC motors with L293D motor driver IC in Case 2.

ii. We can connect a maximum of two motors, and we can rotate both motors in the clockwise and anticlockwise direction. The interfacing of two motors with L293D is shown in Figure 3.36.

3.10.3 Description of L293D Motor Driver Module

The L293D motor driver module is shown in Figure 3.37. The Input 1 (I/P 1), Input 2 (I/P 2), Output 1 (O/P 1), and Output 2 (O/P 2) of the L293D motor driver PCB board are connected to the Pins 2, 7, 3, and 6 of L293D IC, respectively (refer to Figures 3.28 and 3.29 pin diagram of L293D IC). The Input 3 (I/P 3), Input 4 (I/P 4), Output 3 (O/P 3), and Output 4 (O/P 4) of the L293D motor driver PCB board are

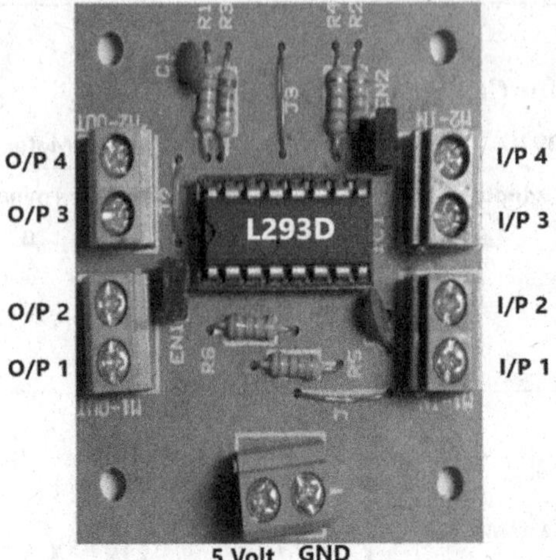

FIGURE 3.37 The L293D motor driver module.

connected to the Pins 10, 15, 11, and 14 of L293D IC, respectively. The 5V and GND (ground) of the L293D motor driver PCB board are connected to the GND pins (Pins 4, 5, 12, and 13) and Vcc pins (Pins 8 and 16), respectively. The Enable 1–2 and Enable 3–4 of L293D motor driver IC are internally connected to Logic 1 on the PCB board; therefore, there is no need to connect Pins 1 and 9. The L293D motor driver's output pins are to be connected to Terminal 1 and Terminal 2 of the DC motor for its operation.

* *Refer to Section 4.11 of Chapter 4 for programming and interfacing of the DC motor with Arduino UNO using motor driver.*

3.11 RELAY

We cannot directly turn on or turn off a heater or AC machine or anything we may think directly from our microcontroller. We need some device which can switch on or off high-power electric devices; these are called relays. Relay is an electromechanical device controlled by small voltage/current and switch on or off high-voltage/current device. Using relay, we can control the AC-operated devices by using a small DC signal.

3.11.1 Pin Description of the Relay

The image of a relay is shown in Figure 3.38a, and the schematic of a relay is shown in Figure 3.38b. A relay is a five-pin device. The five pins of a relay are A, B, COM (common or pole), NC (normally close), and NO (normally open). Terminals A and B are to be connected to low DC voltage for operating the relay. Pins COM (pole), NC, and NO are used to connect high-voltage devices. There is a spring-loaded strip made up of a material that can be attracted by a magnetic field.

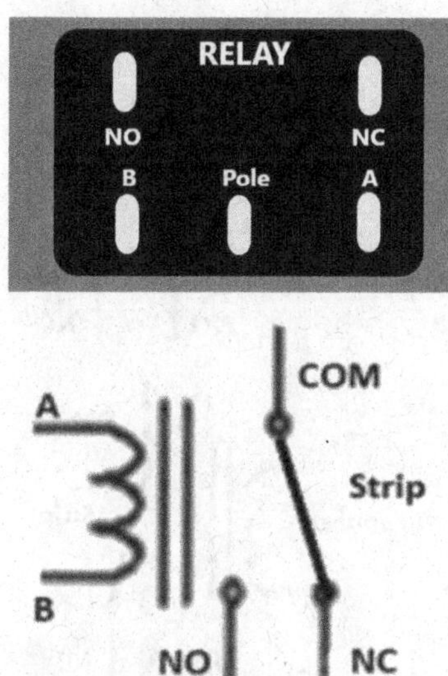

FIGURE 3.38 (a) Image of a relay. (b) Schematic of a relay.

3.11.2 Working of Relay

The schematic of a relay is shown in Figure 3.38b. Pins A and B are to be connected to low voltage to operate the relay. There is a spring-loaded strip made up of a material that can be attracted by a magnetic field. When the relay is not triggered (turn off), COM (pole) and NO terminals are connected via the strip. When the relay is triggered (turn on), COM (pole) and NC terminals are connected via the strip.

The pins A and B of a relay are internally connected to a coil. If a small current passes through this coil (electromagnet), according to Faraday's electromagnetic induction law, a magnetic field will be produced. This magnetic field is sufficient to attract the strip, and thus, COM (pole) terminal will be connected to NO terminal via the strip. This condition is called as triggering of relay or relay is turn on and shown in Figure 3.39 (a).

If we withdraw the small current from A and B pins, then the current conduction through the relay coil will stop, and the magnetic field will vanish. In the absence of a magnetic field, the spring-loaded strip will come back to its normal position; i.e., COM (pole) terminal will be connected to NC terminal via the strip. This condition is called as not triggering of relay or relay is turn OFF and shown in Figure 3.39 (b).

3.11.3 Interfacing of Relay

The switching on/off of high-voltage device will be explained in this section. The interfacing of a relay with high-voltage devices can be done in two ways, as explained below:

Case 1: The interfacing of a relay with a bulb for Case 1 is shown in Figure 3.40 (a). It is evident from the figure that no control voltage is applied between pins A and B, and thus, the relay is not triggered (turn off).

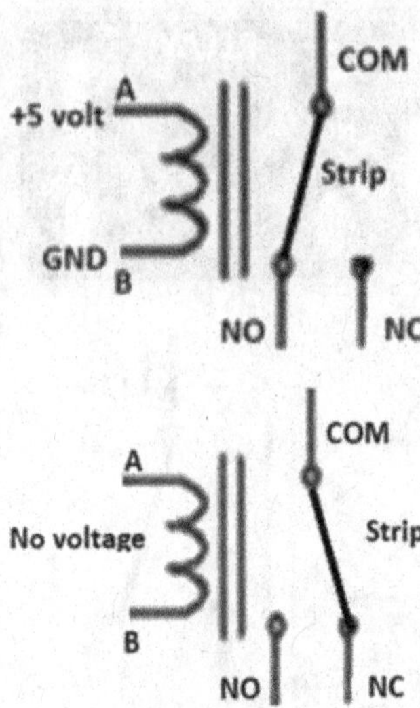

FIGURE 3.39 (a) Relay is triggered (on). (b) Relay is not triggered (off).

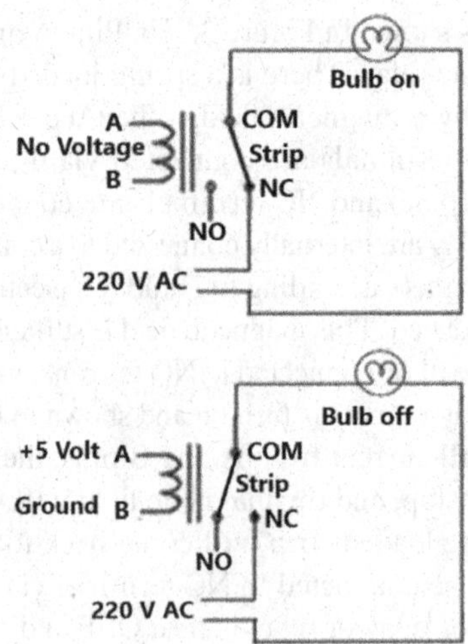

FIGURE 3.40 (a) Relay is not triggered (off) bulb is on. (b) Relay is triggered (on) bulb is off.

Under this condition, COM (pole) terminal is connected to the NC terminal via the strip and the bulb is on due to the short circuit in the bulb conduction path. When the control voltage is applied between pins A and B, as shown in Figure 3.40 (b), the relay will be triggered (turn on). Under this condition, the strip is attracted by the magnetic field produced due to the conduction of current in the internal coil of the relay. The COM (pole) terminal is connected to the NO terminal via the strip, and the bulb is OFF due to the open circuit in bulb conduction path.

Conclusion

The interfacing of a bulb with relay as shown in Case 1 will make bulb on when the relay is not triggered (i.e., relay is turned off) and make bulb off when the relay is triggered (i.e., relay is turned on)

Case 2: The interfacing of a relay with a bulb for Case 2 is shown in Figure 3.41 (a). It is evident from the figure that no control voltage is applied between pins A and B, and thus, the relay is not triggered (turn off). Under this condition, COM (pole) terminal is connected to the NC terminal via the strip and the bulb is off due to the open circuit in the bulb conduction path. When the control voltage is applied between pins A and B, as shown in Figure 3.41(b), the relay will be triggered (turn on). Under this condition, the strip is attracted by the magnetic field produced due to the conduction of current in the relay's internal coil and the COM terminal is connected to NO terminal via the strip and the bulb is on due to the short circuit in the bulb conduction path.

Conclusion

The interfacing of a bulb with relay as shown in Case 2 will make bulb off when the relay is not triggered (i.e., relay is turned off) and make bulb on when the relay is triggered (i.e., relay is turned on).

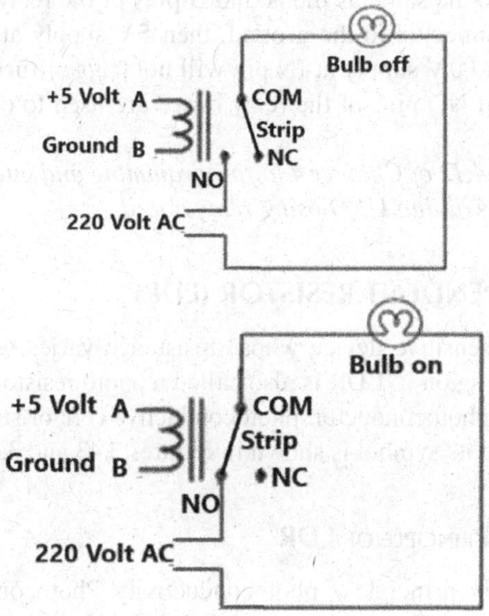

FIGURE 3.41 (a) Relay is not triggered (off) bulb is off. (b) Relay is triggered (on) bulb is on.

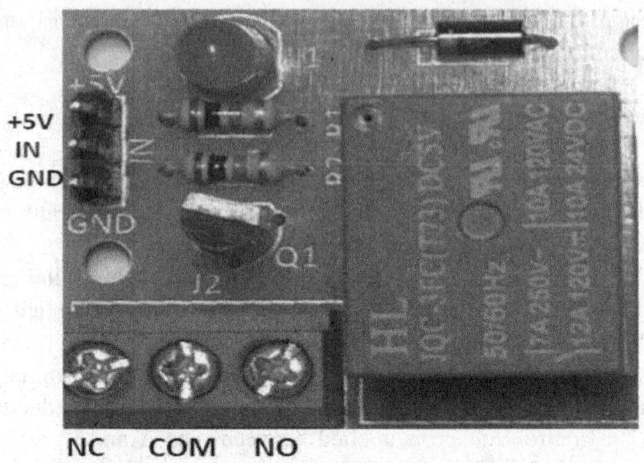

FIGURE 3.42 Relay board module.

3.11.4 Relay Board

The relay board is shown in Figure 3.42. A 5V DC relay is mounted on a PCB along with other components. The LED on the board is for the indication of the on condition of the relay. The +5 V and GND of the relay board should be connected to the 5 V and GND (ground) pin of the Arduino board. This supply voltage will provide an operating voltage to the components mounted on the relay board. If we compare the relay board pins with the relay board's schematic diagram, IN and GND pins of the relay board are the same as the A and B pins of the relay. If the GND pin of the relay board is connected to the ground, then 5 V supply at IN pin will trigger (turn on) the relay and 0 V supply at IN pin will not trigger (turn off) the relay. The NC, COM (pole), and NO pins of the relay board are used to connect high-voltage devices.

* *Refer to Section 4.12 of Chapter 4 for programming and interfacing of the high-voltage devices with Arduino UNO using relay board.*

3.12 LIGHT-DEPENDENT RESISTOR (LDR)

The LDR is a light-sensitive device whose resistance varies with a change in the intensity of light falling on it. LDR is also called a photo resistor. In literature, LDR is also referred to as photoconductor, photoconductive cell, or simply photocell. The image of an LDR and its symbol is shown in Figures 3.43 and 3.44.

3.12.1 Working Principle of LDR

The LDR works on the principle of photoconductivity. Photoconductivity is an optical phenomenon in which the conductivity of material changes when light falls over

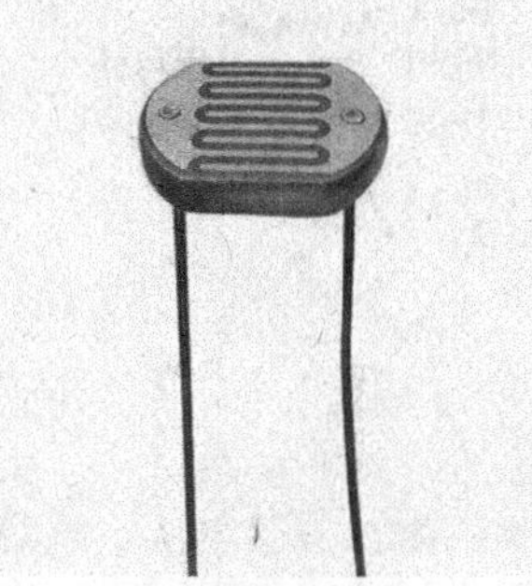

FIGURE 3.43 Image of LDR.

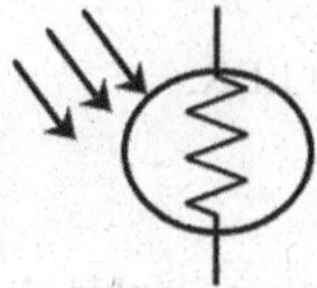

FIGURE 3.44 LDR symbol.

it. The LDR is made up of semiconductor material. In semiconductor materials when light or photons fall, electrons in the valence band are excited to the conduction band provided the photons have energy greater than the forbidden gap of the semiconductor. Thus, falling photons increase the availability of electrons in the conduction band. The increment in the number of electrons in the conduction band increases the device's current flow if it is appropriately connected to the supply. Thus, it is said that the resistance of the device is decreased. The LDRs are cheap, small, and easily available in the market.

Conclusion

The resistance of LDR decreases with the increment in the brightness. The characteristic curve of LDR is shown in Figure 3.45.

3.12.2 Construction of LDR

The structure of LDR is shown in Figure 3.46. The LDR has cadmium sulfide (CdS) film deposited like a track from one side to the other. On the top and bottom, there is a metal film connected to the leads of the LDR. The structure is placed in a transparent plastic package to provide free access to external light. The main component in making LDR is cadmium sulfide (CdS), which acts as a photoconductor and has very little free electrons in darkness. The LDR has manufactured so that its resistance is

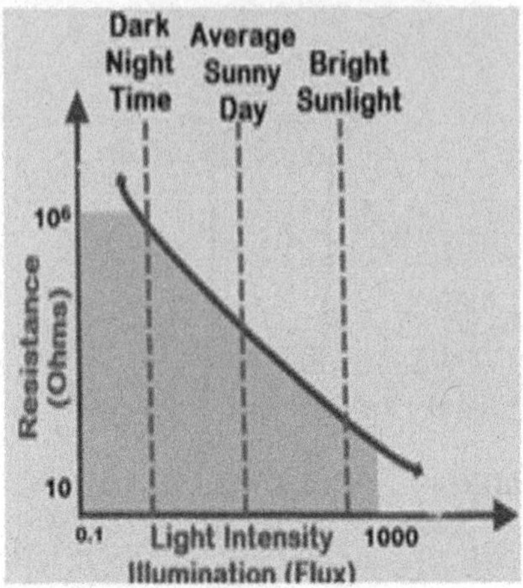

FIGURE 3.45 Characteristic curve of LDR.

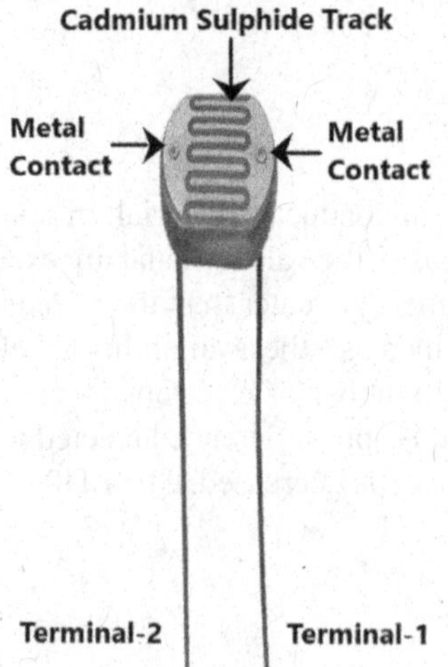

FIGURE 3.46 Structure of LDR.

very high in the range of MΩ in darkness. In the presence of light, the resistance of LDR drops to less than 1 KΩ.

3.12.3 Applications of LDR

The various applications of LDR are listed below:

i. Street light automation
ii. Automation of alarm clock

iii. Light intensity meter
iv. Counting packages moving on a conveyor belt
v. Detection of absence or presence of light.

** Refer to Section 4.13 of Chapter 4 for programming and interfacing of the light-dependent resistor (LDR) with Arduino UNO using relay board.*

3.13 KEYPAD MATRIX

We have already discussed the switches and their interfacing as an input device in Section 3.2. In Section 3.2, we discussed how to interface a single switch with an input port. The situation may arise in a practical situation when we need to interface more than one switch. There are two methods to interface more than one switch with input terminals, and they are the single-dimensional interfacing approach and the two-dimensional interfacing approach.

3.13.1 SINGLE-DIMENSIONAL INTERFACING APPROACH OF SWITCHES

Let us take an example to interface four switches with input port as a single-dimensional interfacing approach as shown in Figure 3.47. The Terminal 2 of each switch is connected to the ground, and Terminal 1 of each switch is connected to 5 V

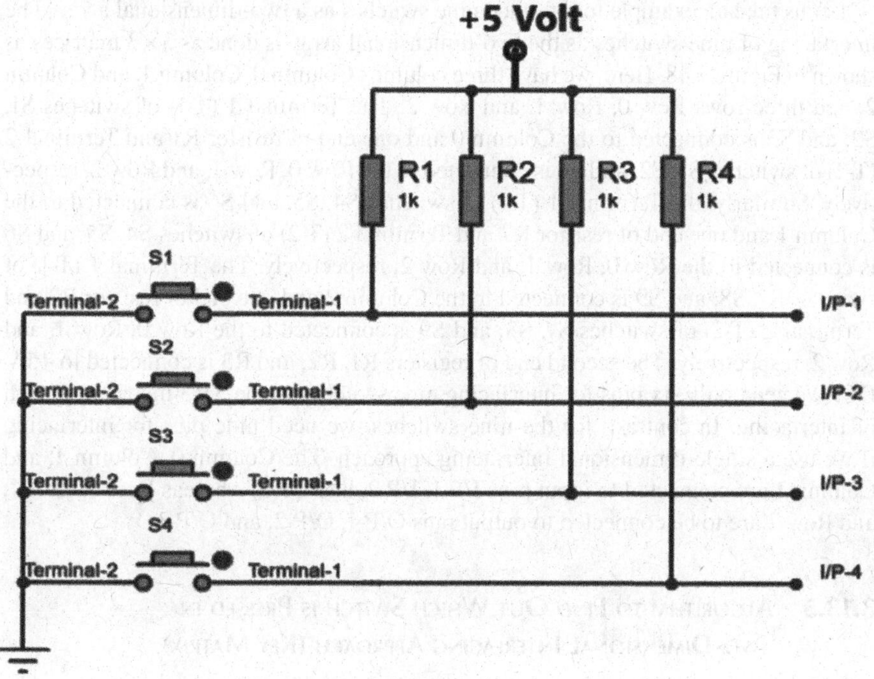

FIGURE 3.47 A single-dimensional interfacing approach of four switches.

through the 1 KΩ register. The junction of Terminal 1 and 1 KΩ register is connected to the input port. The four switches S1, S2, S3, and S4 are connected to the input ports I/P-1, I/P-2, I/P-3, and I/P-4.

3.13.2 Algorithm to Find Out Which Switch is Pressed in Single-Dimensional Interfacing Approach

If no switch is pressed, the input port will read 5 V. The moment a switch is pressed, the corresponding input port will read 0 V. Thus, a 0 V at any input port indicates that the switch attached to the corresponding input port is pressed.

3.13.3 Disadvantage of Single-Dimensional Interfacing Approach

The primary disadvantage of a single-dimensional interfacing approach is that the number of input ports needed for the switches' interfacing is the same as the number of switches. In this approach, if we need to interface more switches, then there may be the possibility that we may exhaust all input ports.

3.13.4 Two-Dimensional Interfacing Approach of Switches

The disadvantage of exhausting all input ports in a single-dimensional interfacing approach of switches can be overcome in two-dimensional interfacing approaches.

Let us take an example to interface nine switches as a two-dimensional array. The interfacing of nine switches as the two-dimensional array is done as 3×3 matrices as shown in Figure 3.48. Here, we have three columns Column 0, Column 1, and Column 2, and three rows Row 0, Row 1, and Row 2. The Terminal 1 (T-1) of switches S1, S2, and S3 is connected to the Column 0 and one end of resistor R1 and Terminal 2 (T-2) of switches S1, S2, and S3 is connected to the Row 0, Row 1, and Row 2, respectively. Similarly, the Terminal 1 (T-1) of switches S4, S5, and S6 is connected to the Column 1 and one end of resistor R2 and Terminal 2 (T-2) of switches S4, S5, and S6 is connected to the Row 0, Row 1, and Row 2, respectively. The Terminal 1 (T-1) of switches S7, S8, and S9 is connected to the Column 2 and one end of resistor R3 and Terminal 2 (T-2) of switches S7, S8, and S9 is connected to the Row 0, Row 1, and Row 2, respectively. The second end of registers R1, R2, and R3 is connected to +5 V DC. We need only six pins for interfacing nine switches in the 3×3 matrix approach of interfacing. In contrast, for the nine switches, we need nine pins for interfacing if we use a single-dimensional interfacing approach. The Column 0, Column 1, and Column 2 are connected to input pins I/P-1, I/P-2, and I/P-3, whereas Row 0, Row 1, and Row 2 are to be connected to output pins O/P-1, O/P-2, and O/P-3.

3.13.5 Algorithm to Find Out Which Switch is Pressed in Two-Dimensional Interfacing Approach (Key Matrix)

During the interfacing of a key matrix, we must ensure that all the row lines must be connected to the output pins and all the column lines must be connected to

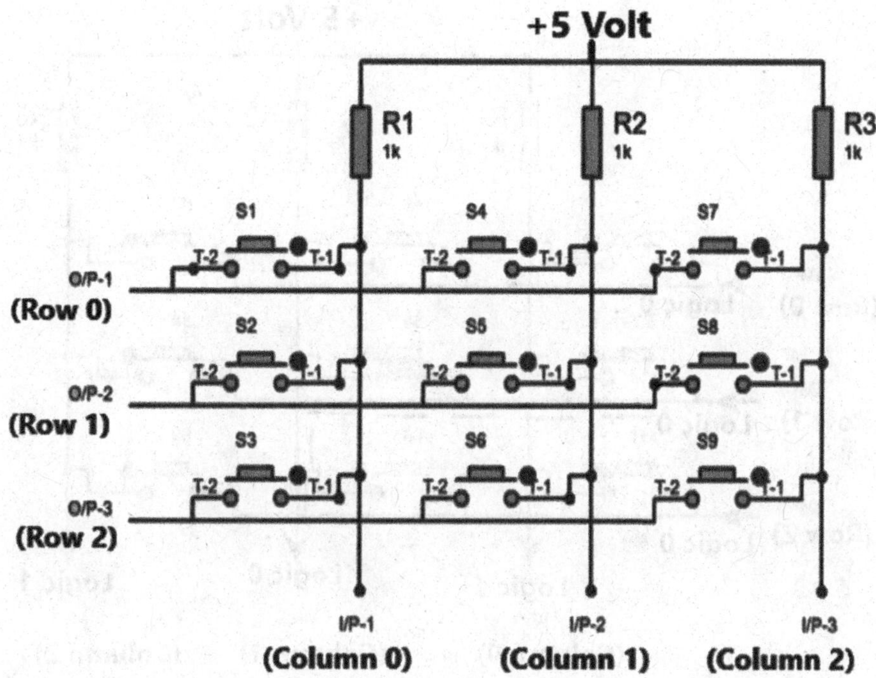

FIGURE 3.48 A two-dimensional interfacing approach of nine switches.

the input pins. To discuss the algorithm to find out which switch is pressed in the two-dimensional interfacing approach (key matrix), let us take a 3×3 key matrix as shown in Figure 3.48.

Step 1: Send Logic 0 (0 V) to all output pins O/P-1, O/P-2, and O/P-3.

Step 2: Scan the input pins I/P-1, I/P-2, and I/P-3 sequentially and continuously. Let us consider when no key is pressed, all the input pins will be connected to 5 V via registers R1, R2, and R3 and input pins read Logic 1 (5 V).

Step 3: Let us consider when a key is pressed; say, for example, S5 is pressed as shown in Figure 3.49. In this case, the I/P-1 and I/P-3 input pins will be connected to 5 V via registers R1 and R3, respectively, but the I/P-2 input pin will be connected to Logic 0 (0 V) through the short-circuit path created by the pressed S5 key. The readers should understand that if any switch in Column 1 (S4, S5, and S6) is pressed, I/P-2 will be at Logic 0 (0 V). Thus, a Logic 0 (0 V) at I/P-2 pin indicates that a switch in Column 1 (S4, S5, and S6) is pressed. Similarly, if any switch in Column 0 (S1, S2, and S3) is pressed, I/P-1 will be at Logic 0 (0 V). Thus, a Logic 0 (0 V) at I/P-1 pin indicates that a switch in Column 0 (S1, S2, and S3) is pressed. Similarly, if any switch in Column 2 (S7, S8, and S9) is pressed, I/P-3 will be at Logic 0 (0 V). Thus, a Logic 0 (0 V) at I/P-3 pin indicates that a switch in Column 2 (S7, S8, and S9) is pressed. So in Step 3, we can get an indication that a switch in a specific column is pressed, but still, it is not clear which switch is pressed exactly.

Step 4: We shall continue with the same example in which the S5 switch is pressed. In Step 3, we identified that a switch in Column 1 is pressed. To identify the key of

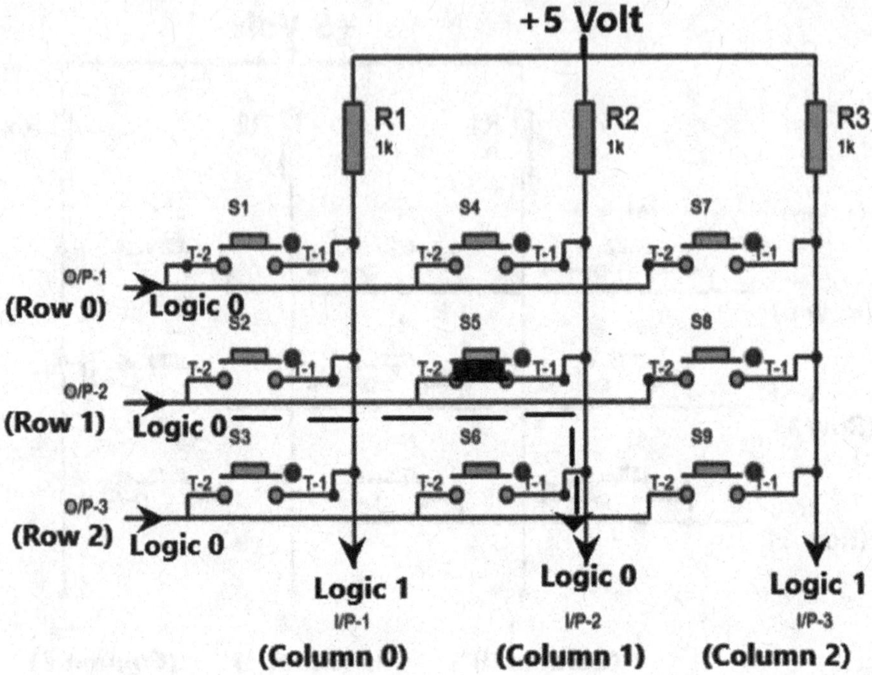

FIGURE 3.49 Illustration of the logic level available at three columns when switch S5 is pressed in a 3×3 key matrix.

which row is pressed, we first send Logic 0 (0 V) at O/P-1 output pin and Logic 1 at O/P-2 and O/P-3 pins. If we scan the input pins, we get the I/P-1 pin at Logic 1, I/P-2 pin at Logic 1, and I/P-3 pin at Logic 1 (5 V). The Logic 1 (5 V) at all input pins is the indication that the key at Row 0 of Column 1, i.e., S4, is not pressed. Now we send Logic 0 (0 V) at O/P-2 output pin and Logic 1 at O/P-1 and O/P-3 pins. If we scan the input pins, we get I/P-1 pin at Logic 1 (5 V), I/P-2 pin at Logic 0, and I/P-3 pin at Logic 1 (5 V). The Logic 0 (0 V) at I/P-2 pin is the indication that the key at Row 1 of Column 1, i.e., S5, is pressed. Suppose we send Logic 0 (0 V) at O/P-3 output pin and Logic 1 at O/P-1 and O/P-2 pins. If we scan the input pins, we get all the input pins I/P-1, I/P-2, and I/P-3 at Logic 1 (5 V). The Logic 1 (5 V) at all input pins is the indication that the key at Row 2 of Column 1, i.e., S6, is not pressed.

3.13.6 A 4×4 Keypad

A 4×4 keypad available in the market is shown in Figure 3.50. A 4×4 keypad has four rows and four columns. Pins 1, 2, 3, and 4 are the pins for Row 0, Row 1, Row 2, and Row 3, respectively, and Pins 5, 6, 7, and 8 are the pins for Column 0, Column 1, Column 2, and Column 3, respectively. The position of various keys in the keypad with respect to row and column is shown in Table 3.10.

** Refer to Section 4.14 of Chapter 4 for programming and interfacing of the keypad with Arduino UNO.*

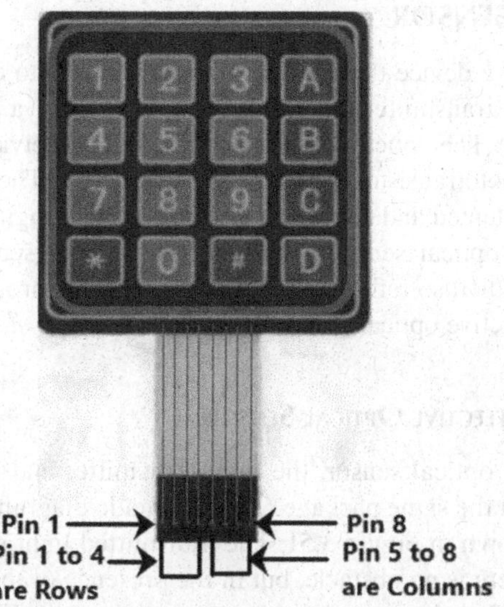

FIGURE 3.50 A 4×4 keypad.

TABLE 3.10
The Position of Various Keys in the Keypad with Respect to Row and Column

Key Name	Row Number	Column Number
1	0	0
2	0	1
3	0	2
A	0	3
4	1	0
5	1	1
6	1	2
B	1	3
7	2	0
8	2	1
9	2	2
C	2	3
*	3	0
0	3	1
#	3	2
D	3	3

3.14 OPTICAL SENSOR

An optical sensor is a device that converts the light rays into electrical signals. An optical sensor has a transmitter that emits light. Generally, a LED or a LASER is used as a transmitter. The optical sensor also has a light-receiving device. Generally, a photodiode or a phototransistor is used as a light receiver. The light received by the optical sensor is evaluated and converted into an electrical signal.

There are three optical sensors: through-bean optical sensors, retro-reflective optical sensors, and diffuse-reflective optical sensors. The through-bean optical sensor and diffuse-reflective optical sensor are out of the scope of this book.

3.14.1 Retro-Reflective Optical Sensor

In a retro-reflective optical sensor, the light transmitter and light receiver – both units – are placed in the same package. The schematic diagram of a retro-reflective optical sensor is shown in Figure 3.51. The transmitted light will never come back to the receiver if there is no obstacle, but in the presence of the obstacle, the transmitted light will reflect back to the receiver, as shown in Figure 3.51. Once the receiver receives the light, it is evaluated by the photodiode or a phototransistor and converted into an electrical signal. The symbol of a bi-terminal phototransistor is shown in Figure 3.52. The two terminals of the phototransistor are the emitter (with arrow mark) and the collector. The phototransistors are specially designed and optimized for light sensitivity. In the symbol of the phototransistor, two arrows

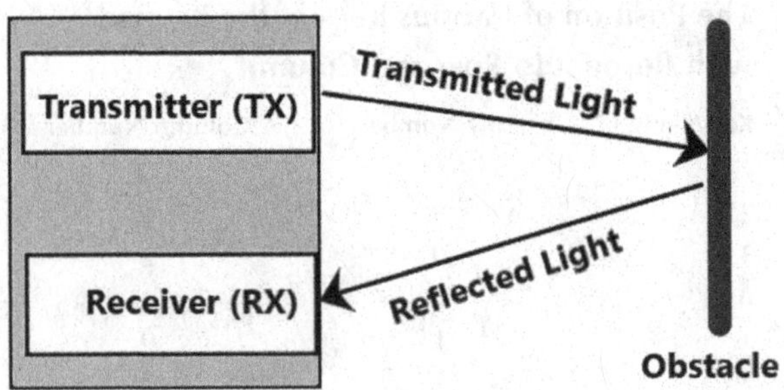

FIGURE 3.51 A schematic of retro-reflective optical sensor.

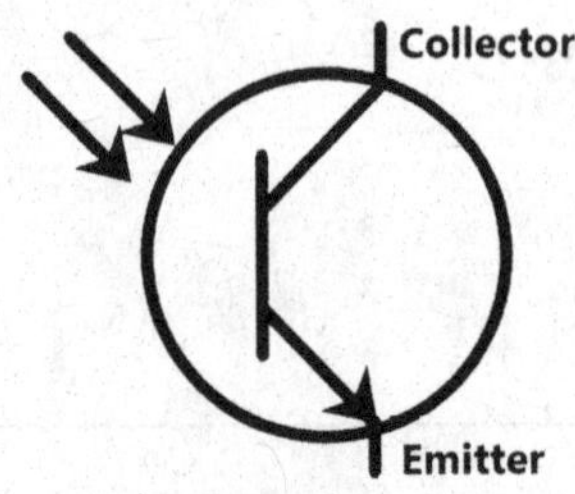

FIGURE 3.52 A symbol of phototransistor.

pointing towards the base indicate that the device is sensitive to light. A phototransistor circuit is shown in Figure 3.53, and two cases can explain its operation.

Case I: When there is no obstacle, no transmitted light will be reflected back, and no light will fall on the phototransistor. Thus, the phototransistor is in cut-off condition. No current flows in the collector to emitter terminal, and collector–emitter terminals will act as an open circuit, and we get Vout = 5 V. The working principle of phototransistor when there is no obstacle is shown in Figure 3.54.

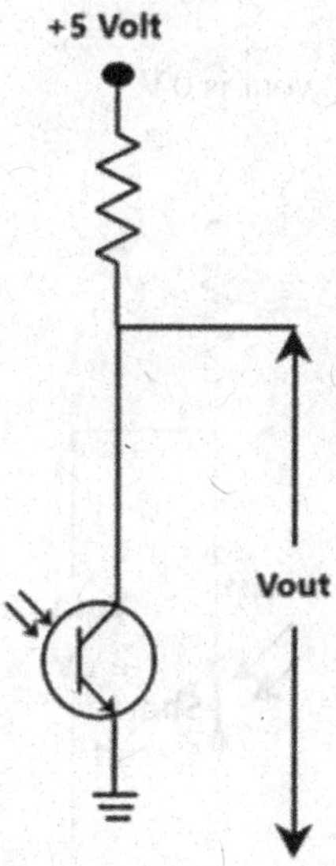

FIGURE 3.53 The phototransistor circuit.

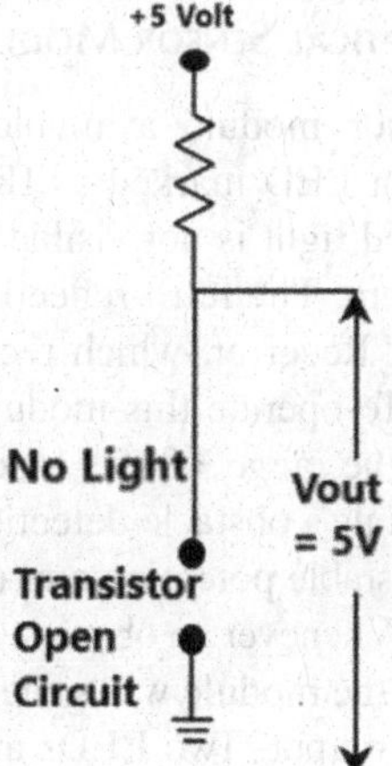

FIGURE 3.54 Case I – Working principle of phototransistor when there is no obstacle.

Conclusion

In the absence of an obstacle, Vout is 5 V.

Case II: When there is an obstacle, transmitted light will be reflected back and fall on the phototransistor. Thus, phototransistor conducts and current flows in the collector to emitter terminal, and collector–emitter terminals will act as short circuits and get Vout = 0 V. The working principle of phototransistors when there is an obstacle is shown in Figure 3.55.

Conclusion

In the presence of an obstacle, Vout is 0 V.

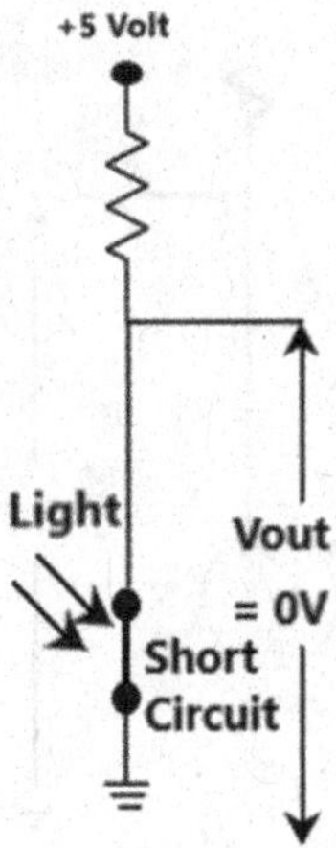

FIGURE 3.55 Case II – Working principle of phototransistor when there is an obstacle.

3.14.2 Retro-Reflective Optical Sensor Module

A retro-reflective optical sensor module available in the market is shown in Figure 3.56. It has a transparent LED marked as IR Emitter, which emits light in the infrared region. The infrared light is not visible by naked eyes, but we can see this light from the mobile camera. The retro-reflective optical sensor module has a black color LED marked as IR Receiver, which receives the infrared light, which is reflected from the obstacle. To operate this module, we have to connect the Vcc marked pin to a DC voltage in the range 3.3–5 V, and Gnd marked pin is to be connected to the ground. This module's obstacle detection range is 2–30 cm, and it can be adjusted by rotating the adjustable potentiometer (marked as distance potentiometer on optical sensor module). Whenever an obstacle comes in front of the IR emitter LED, the pin marked as Out on the module will generate 0 V output, and when there is no obstacle, it generates 5 V output. Two LEDs are also mounted on the optical sensor module. The first LED is the power LED, and it is turned on when the module is appropriately powered on by connecting Vcc and Gnd pins at specified voltage

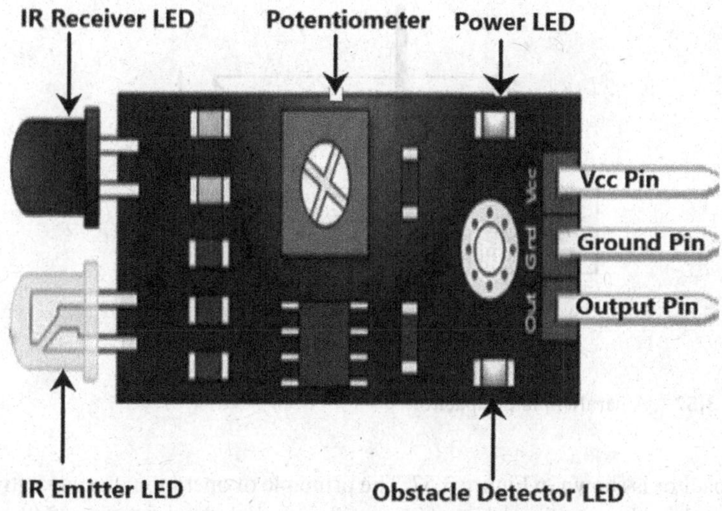

FIGURE 3.56 A retro-reflective optical sensor module.

levels. The second LED is an obstacle LED, and it is turned on when the obstacle comes in front of the module; otherwise, it is off.

The technical specifications of the retro-reflective optical sensor module are as follows:

i. Operation voltage: 3.3–5 V
ii. Active output level: outputs low logic level when an obstacle is detected
iii. Detection range: 2–30 cm (adjustable using potentiometer)
iv. Vcc (input pin): 3.3–5 V power input
v. Gnd: 0 V
vi. Out: digital output pin.

* *Refer to Section 4.15 of Chapter 4 for programming and interfacing of the optical sensor with Arduino UNO.*

3.15 CAPACITIVE TOUCH SENSOR

A capacitive touch sensor works on the principle of capacitive effect, which can detect the physical touch. We are all familiar with many of the touch-activated devices; for example, our mobiles are touch-sensitive. We use our fingers to navigate on the screen. There are many other input devices where we can touch the screen and feed our inputs. By using the capacitive touch-sensing technology, we can implement touch-activated types of the sensing device.

We already had some discussions on capacitance in Section 3.9, and for the benefit of the readers, we shall reproduce them here. A capacitive sensor converts a non-electrical quantity (e.g., force, pressure, humidity) into an electrical quantity (e.g., voltage or current) using the capacitance change. The schematic of a parallel

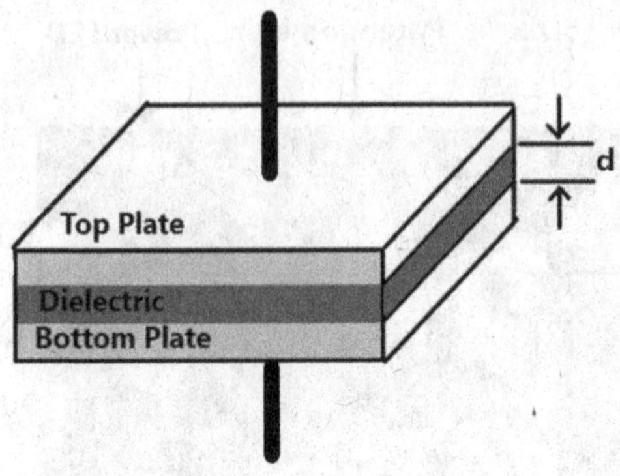

FIGURE 3.57 A parallel plate capacitor.

plate capacitor is shown in Figure 3.57. The principle of operation of a capacitive sensor is based on the equation of capacitance of a parallel plate capacitor (3.8).

$$C = \varepsilon_r \varepsilon_o A/d \tag{3.8}$$

Here, C = Capacitance of capacitor
ε_r = Relative permittivity
ε_o = Permittivity of free space = 8.85×10^{-12} Farad/meter
A = Area of overlapping parallel plate (m^2)
D = Distance between two parallel plates (m)

The capaeitance of the capacitor will change if there is a change in the area of overlapping parallel plate or distance between two parallel plates or changes in permittivity.

Due to the capacitance change, the capacitive reactance (Xc) of the capacitance will change. The capacitive reactance (Xc) of the capacitance is given by (3.9):

$$Xc = 1/2\pi fC \tag{3.9}$$

Here, Xc = Capacitive reactance (Ω)
f = Frequency of signal (Hz)
C = Capacitance of capacitor (Farad)

The change in voltage level can recognize the change in the capacitive reactance.

3.15.1 Capacitive Touch Sensor Working Principle

Capacitive sensing is based on capacitive coupling. Our body is also conductive. When we bring our finger close to a material, our finger and that material will form a capacitance because some moisture in our finger acts as a dielectric. If we have proper circuitry to detect the capacitance change by bringing our finger close to the conductive plate, we can detect the touch. The operation of the capacitive touch sensor can be explained in two cases.

Case I: When we have not touched a parallel plate capacitive touch sensor, the capacitance of a touch sensor is given by (3.8) and denoted by C0 as shown in Figure 3.58.

Case II: If we touch the top plate of a capacitor with our finger, then a capacitance Cf will be formed between regions of the top plate where we touched and finger as shown in Figure 3.59. When we are approaching our finger near the capacitive touch sensor pad, the distance between the touchpad and our figure decreases, which causes the increment in finger capacitance (Cf) as per (3.8). Since C0 and Cf are connected in parallel, the increment in Cf will cause an increment in effective capacitance (Ce) as Ce = C0 + Cf. As per (3.9), the increment in capacitance will decrease the capacitive reactance (Xc), and thus, the current flow will increase. The electronic circuitry can be designed to detect this current flow or the touch.

3.15.2 Capacitive Touch Sensor Module

A capacitive touch sensor module available in the market is shown in Figure 3.60. It is a three-pin module. The Vcc pin is to be connected to 2.0–5.5 V. The Gnd pin is to be connected to the ground. The digital output pin generates 0 or 5 V depending upon the capacitive touch sensor's touch or no touch. If we touch the sensor's touchpad, then the module's digital output pin generates 5 V (Logic 1); otherwise, it generates 0 V (Logic 0). The response time of this sensor is 60–220 ms. The signal conditioning IC TTP223B is used to convert the touch sensor's change in capacitance into an electrical signal.

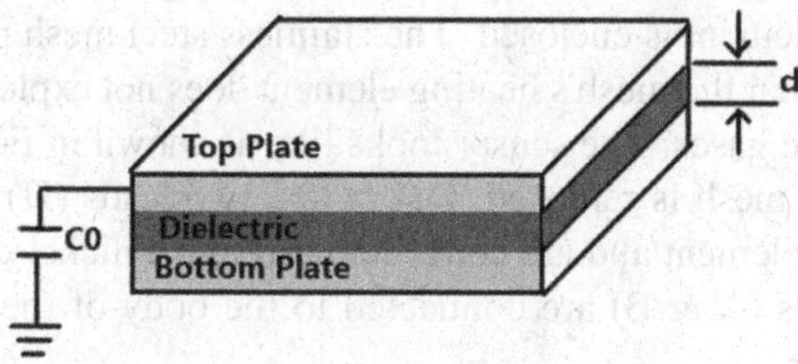

FIGURE 3.58 The parallel plate capacitor when not touched.

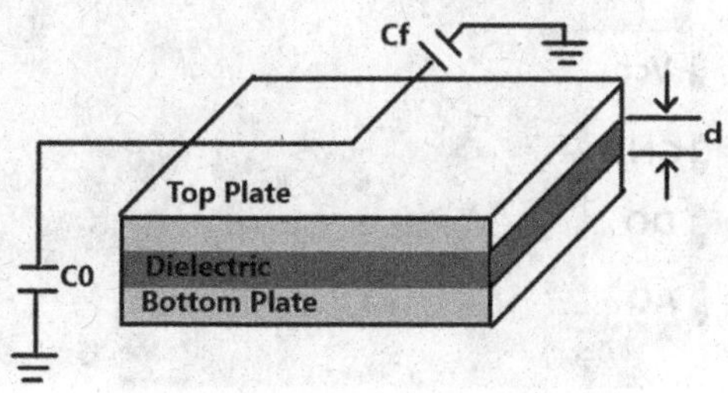

FIGURE 3.59 The parallel plate capacitor when touched.

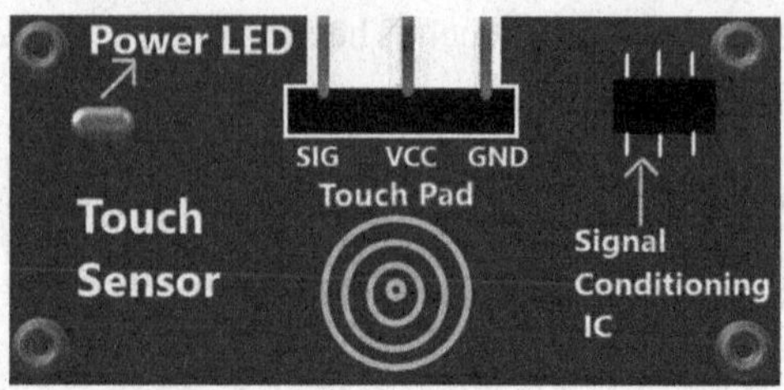

FIGURE 3.60 A capacitive touch sensor module.

** Refer to Section 4.16 of Chapter 4 for programming and interfacing of the touch sensor with Arduino UNO.*

3.16 GAS SENSOR

Gas sensor module (MQ2) can be used for sensing LPG, smoke, alcohol, propane, hydrogen, methane, and carbon monoxide concentrations in the air anywhere between 200 and 10,000 ppm. It is a semiconductor-based gas sensor in which the gas detection is carried out by a change of resistance of the sensing material when the gas comes in contact with the sensor.[3]

Parts per million is abbreviated as ppm, and it is the ratio of one gas to another. For example, 1,000 ppm of carbon monoxide means that if you could count a million gas molecules, 1,000 of them would be carbon monoxide and 999,000 molecules would be some other gases.[3]

The top view and bottom view of the MQ2 gas sensor are shown in Figures 3.61 and 3.62. The MQ2 gas sensor's top view is showing stainless steel mesh inside which the sensing element is enclosed. The stainless steel mesh protects the sensing element. It ensures that the mesh's heating element does not explode when the sensor is sensing flammable gases. The sensor looks like as shown in Figure 3.63 when the outer stainless steel mesh is removed. Out of six, two leads (H) are responsible for heating the sensing element and are connected through a nickel-chromium coil. The remaining four leads (A & B) are connected to the body of the sensing element.[3]

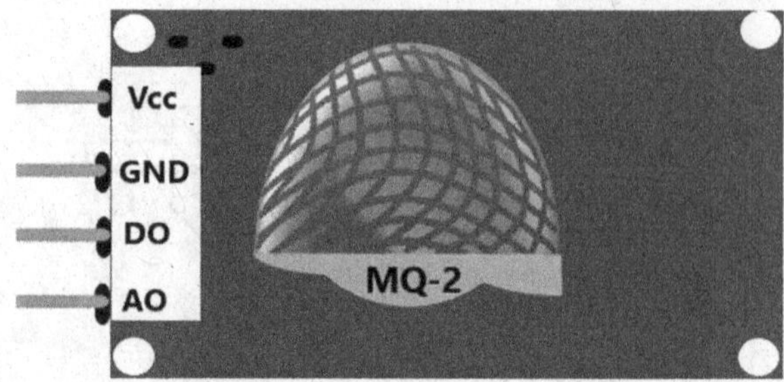

FIGURE 3.61 Top view of MQ2 gas sensor.

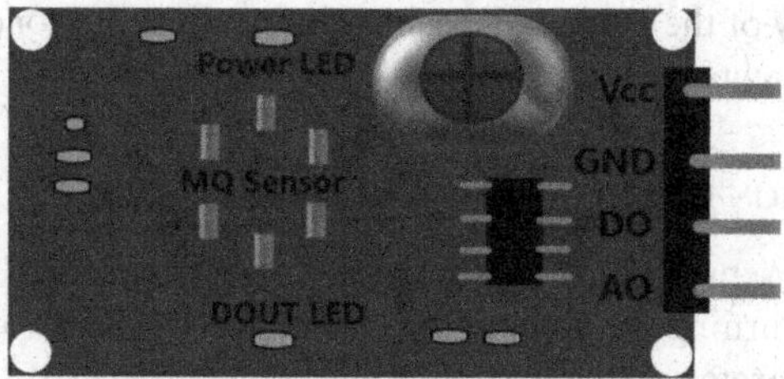

FIGURE 3.62 Bottom view of MQ2 gas sensor.

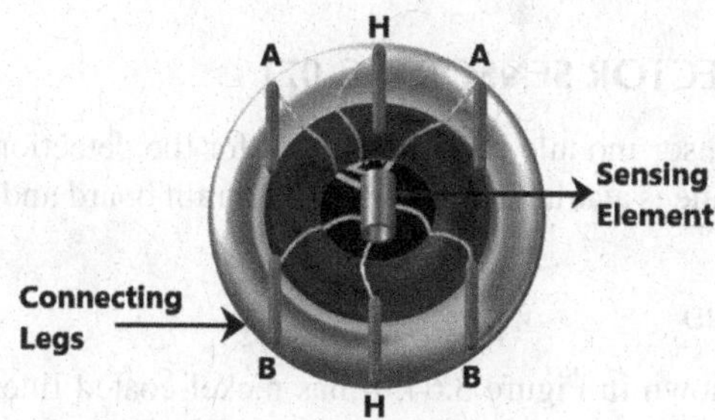

FIGURE 3.63 The sensing element when stainless steel mesh is removed.

These four leads (A & B) are responsible for output signals. The sensing element is made up of aluminum oxide (AL_2O_3) with a coating of tin dioxide (SnO_2). Tin dioxide is the most critical material being sensitive to combustible gases.

3.16.1 Working of the Gas Sensor (MQ2)

When tin dioxide is heated in air, oxygen is adsorbed on the surface. In clean air, donor electrons in tin dioxide are attracted to oxygen that is adsorbed on the sensing material's surface. This process decreases the number of free electrons in tin dioxide and thus prevents electric current flow. If the surrounding space has combustible gases, then oxygen reacts with these gases and releases the donor electrons of tin dioxide previously attracted to oxygen. Thus, the current flow in the sensor suddenly increases.

3.16.2 Gas Sensor Module (MQ2)

The bottom view of the MQ2 gas sensor is shown in Figure 3.62. To operate this module, we have to connect the Vcc marked pin to 5 V and GND marked pin is to be connected to the ground. A power LED turns on when the module is appropriately powered on by connecting Vcc and GND pins at specified voltage levels.

The gas sensor (MQ2) module comes with two different outputs, i.e., digital output (DO) and analog output (AO). The analog output voltage provided by the sensor changes is proportional to the concentration of gas. The potentiometer can be used to adjust the sensitivity of the sensor. We can use it to adjust the concentration of gas at which the sensor detects it.

3.16.3 Calibration of Gas Sensor (MQ2) Module

To calibrate the gas sensor, we should bring it in the proximity of the gas we want to detect and keep on turning the potentiometer until the Red LED on the module starts glowing. The comparator on the module continuously checks if the analog pin (A0) has hit the potentiometer's threshold value. When it crosses the threshold, the digital pin (D0) will go high, and the signal LED turns on.[3]

* *Refer to Section 4.176 of Chapter 4 for programming and interfacing of the gas sensor with Arduino UNO.*

3.17 RAIN DETECTOR SENSOR (FC-07)

The rain detector sensor module (FC-07) is used for the detection of rain. The rain detector sensor module is available in two boards: a rain board and the control board.

3.17.1 Rain Board

The rain board is shown in Figure 3.64. It has nickel-coated lines, and it works on the principle of resistance. It shows more resistance when it is dry and less resistance when it is wet.[8]

3.17.2 Rain Sensor Control Board

The rain sensor control board is shown in Figure 3.65. The control board has four pins on one side and two pins on another side. To operate this module, we have to connect the VCC marked pin to 5 V, and GND marked pin is to be connected to the ground. The rain sensor module comes with two different outputs, i.e., digital output and analog output. The analog output voltage provided by the sensor changes is proportional to the intensity of water falling on the rain board. The digital output will be high when the water falling on the rain board exceeds the potentiometer's sensitivity level. Power LED turns on when the module is appropriately powered on by connecting VCC and GND pins at specified voltage levels. The potentiometer can be used to adjust the sensitivity of the rain sensor. We can use it to adjust the intensity of water it detects. Suppose we rotate the potentiometer to a clockwise direction. In that case, the sensor becomes more

FIGURE 3.64 The rain board of rain detector sensor.

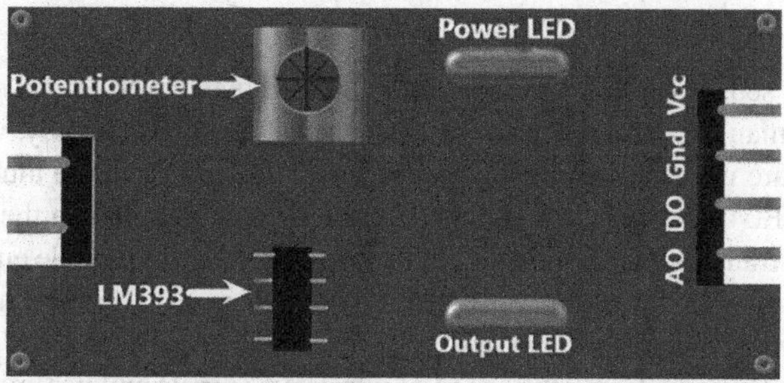

FIGURE 3.65 The control module of rain detector sensor.

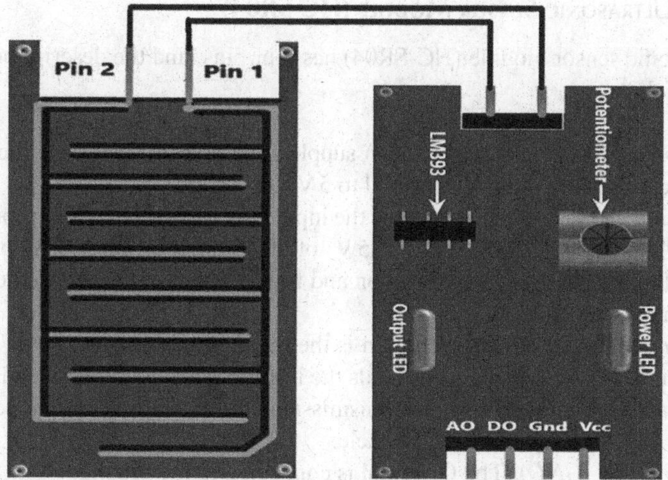

FIGURE 3.66 The rain board and control board of rain detector sensor.

sensitive, and if we rotate the potentiometer to an anti-clockwise direction, the sensor becomes less sensitive. When the water is detected, the output LED turns on. The onboard LM393 comparator IC on the control board is used for signal conditioning.

3.17.3 Working of Rain Detector Sensor (FC-07)

The two pins of the control board available on another side of the board are connected to the two pins of the rain board. These pins can be connected to the rain board in any direction. When there is no water drop on the rain board, its resistance is high, and we get high voltage at the output. When water drops on the rain board, its resistance decreases because water is the conductor of electricity. The presence of water connects nickel lines in parallel, reducing the resistance and reducing the output voltage. The rain board and control board are connected, as shown in Figure 3.66.

* *Refer Section 4.18 of Chapter 4 for programming and interfacing of the gas sensor with Arduino UNO.*

3.18 ULTRASONIC SENSOR (HC-SR04)

The ultrasonic sensor (HC-SR04) is used to find out how much far an object is from the sensor. Human ears can listen to the frequencies up to 20kHz only, whereas those above 20kHz are ultrasonic frequencies. The front-side image of the ultrasonic sensor module (HC-SR04) is shown in Figure 3.67. It can detect an object in the range from 2 to 450cm. The transmitter section (TX) of the ultrasonic sensor module emits ultrasonic signals which travel in the air and is reflected back if there is an obstacle in its path. The reflected ultrasonic signals are captured by the receiver section (RX) of the ultrasonic sensor module. By considering the speed of ultrasonic signals and its time, the Arduino will calculate an object's distance from the sensor by the applicable statement.

3.18.1 Pin Description and Other Details of Ultrasonic Sensor Module (HC-SR04)

The ultrasonic sensor module (HC-SR04) has four pins, and the description of these pins is as follows:

Pin Number 1 (VCC): The VCC pin supplies the power to generate the ultrasonic pulses. It is to be connected to 5V.

Pin Number 2 (Trig): The Trig pin is the input pin of the ultrasonic sensor module (HC-SR04). A HIGH pulse (5V) of 10 μs duration must be applied to the Trig pin to start the generation and transmission of the ultrasonic pulse from the transmitter.

Pin Number 3 (Echo): The Echo pin is the output pin of the ultrasonic sensor module (HC-SR04). This pin sends the information about the time taken by the ultrasonic signals from its transmission point. It captures by the receiver after it gets reflected by the obstacle.

Pin Number 4 (GND): The GND pin is connected to the ground.

The backside image of the ultrasonic sensor module (HC-SR04) is shown in Figure 3.68. The IC MAX3232 can be seen behind the TX of the HC-SR04 module, and it is used to control the transmission of ultrasonic frequency signals. The IC LM324 can be seen behind the RX of the HC-SR04 module. It is a quad Op-Amp that amplifies the signal generated by the receiving section up to a sufficient level to be recognized by the Arduino.

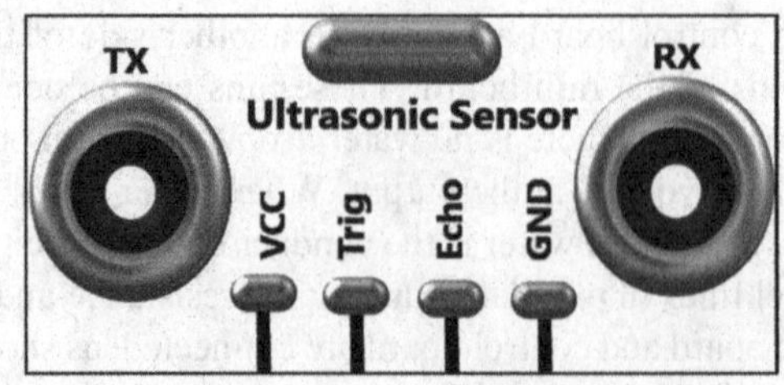

FIGURE 3.67 The ultrasonic sensor module (HC-SR04) – Front-side image.

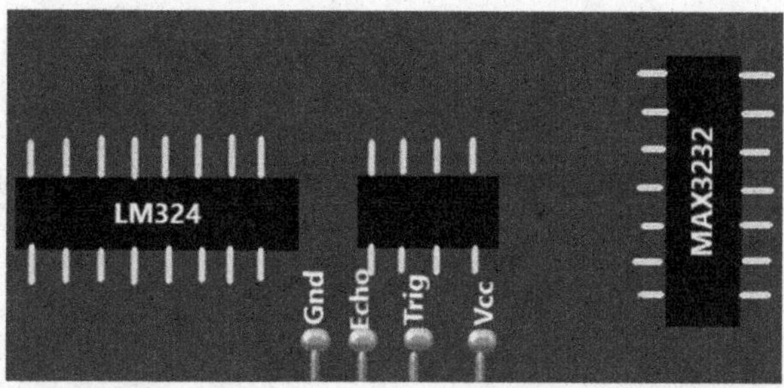

FIGURE 3.68 The ultrasonic sensor module (HC-SR04) – Backside image.

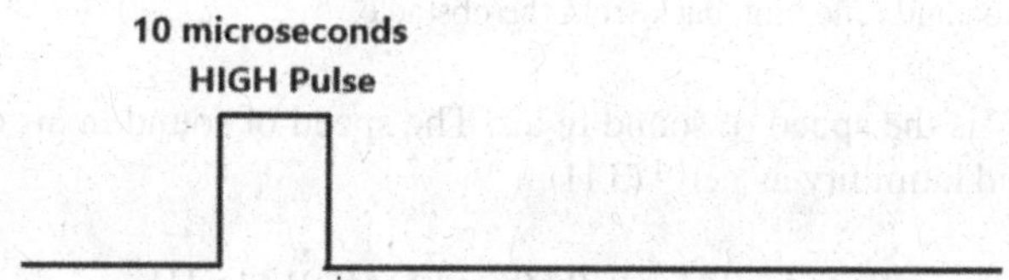

FIGURE 3.69 A 10 μs high pulse to be applied at Trig input pin of the ultrasonic sensor module.

3.18.2 Working Principle of Ultrasonic Sensor Module (HC-SR04)

To start the distance measurement, we need to send a high pulse (5 V) of 10 μs duration to the Trig pin of the sensor module, as shown in Figure 3.69. When the ultrasonic sensor module receives the high signal on the Trig pin, the TX of the sensor module will emit eight ultrasonic pulses of 40 kHz frequency as shown in Figure 3.70. If an object is within range, the eight pulses will be reflected back and received by the sensor module's RX. When the reflected ultrasonic pulse hits the receiving section of the sensor module, the Echo pin outputs a high-voltage signal. The Echo pin output will remain high for the time taken by the eight pulses to travel from the transmitting section and back to the receiving section after getting reflected from the obstacle as shown in Figure 3.71. The relation between the speed of sound, distance, and time traveled is given by (3.10):

$$\text{Distance} = \text{Speed} \times \text{Time} \tag{3.10}$$

The "time" is the time taken by the ultrasonic pulse to leave the transmission section of the HC-SR04 sensor, reflected back from the object, and return to the RX of the sensor module.

However, we only want to measure the distance to the object, not the distance of the path the ultrasonic pulse took. Therefore, we divide that time in half to get the time variable (3.10).

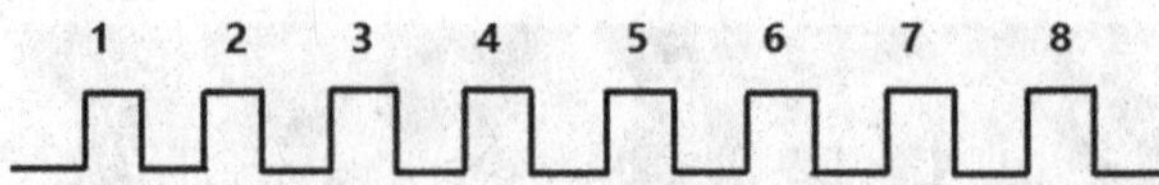

FIGURE 3.70 Eight ultrasonic pulses generated by the transmitter section of the ultrasonic sensor module.

FIGURE 3.71 The high pulse generated by the ECHO output pin of the ultrasonic sensor module is equivalent to the time taken by the ultrasonic pulse to return to the receiver section after its generation and reflection back from the obstacle.

The "speed" is the speed of sound in air. The speed of sound in air changes with temperature and humidity as per[9] (3.11):

$$\text{Speed} = 331.4 + (0.606 \times \text{T}) + (0.0124 \times \text{H}) \tag{3.11}$$

where

Speed of sound in air is in m/s
331.4 m/s = Speed of sound in air at 0°C and 0% humidity
T = Temperature in 0°C
H = % of Relative humidity

Therefore, to accurately calculate the distance of sound in the air, we must consider the ambient temperature and humidity. Still, for simplicity, the speed of sound in air is considered as 340 m/s.

Example 3.6

An object is placed 20 cm away from the ultrasonic sensor module (HC-SR04). Explain the steps involved in calculating the distance of the object from the ultrasonic sensor module (HC-SR04).

Solution

Steps involved in the calculation of the distance of the object from the ultrasonic sensor module (HC-SR04) with reference to Figure 3.72 are as follows:

Step 1: To initiate the process to measure the object's distance from the ultrasonic sensor module, a high pulse (5 V) of 10 μs duration is applied at the Trig pin of the sensor module as shown in Figure 3.69.

Step 2: When the ultrasonic sensor module receives the high signal on the Trig pin, the TX of the sensor module will emit eight ultrasonic pulses of 40 kHz frequency as shown in Figure 3.70.

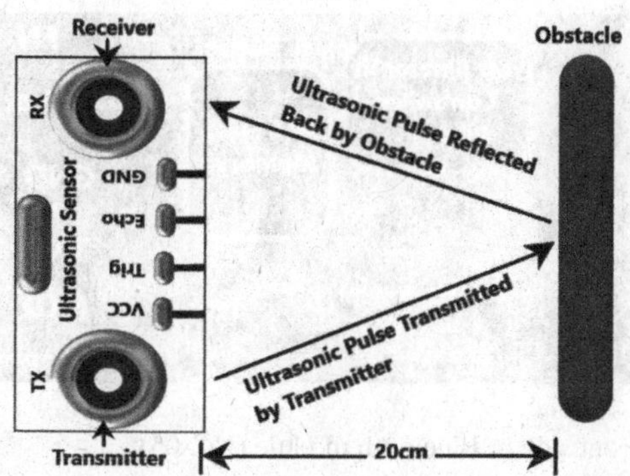

FIGURE 3.72 The ultrasonic sensor module (HC-SR04) and obstacle.

Step 3: The speed of the ultrasonic pulse in the air is 340 m/s or 0.034 cm/μs. Since the object is 20 cm away from the ultrasonic sensor module, ultrasonic pulse requires 588 μs to reach the object. After getting reflected from the object again, the ultrasonic pulse requires 588 μs to reach the sensor module's RX.

Step 4: When the reflected ultrasonic pulse hits the receiving section of the sensor module, the sensor module's Echo pin outputs a high-voltage signal. The Echo pin output will remain high for the time taken by the ultrasonic pulse to travel from the transmitting section and back to the receiving section after getting reflected. Thus, the Echo pin output will remain high for 1,176 μs.

Step 5: Since we are concerned with the time taken by the ultrasonic pulse to hit the object, we divide the time duration for which the Echo pin output is high by 2. When the object is 20 cm away from the ultrasonic sensor module (HC-SR04), the time taken by the ultrasonic pulse to hit the object will be 588 μs.

Step 6: We know from (3.10) that Distance = Speed × Time.

Put the value speed of sound in air which is considered as 0.034 cm/μs and time 588 μs in (3.10), then distance will be calculated as 20 cm.

* *Refer to Section 4.19 of Chapter 4 for programming and interfacing of the ultrasonic sensor with Arduino UNO.*

3.19 BLUETOOTH MODULE (HC-05)

The front-side and backside images of the Bluetooth module (HC-05) are shown in Figures 3.73 and 3.74, respectively. The Bluetooth module works in ISM band frequency from 2.4 to 2.485 GHz. The name of "Bluetooth" is kept in tenth-century king Harald Bluetooth who united dissonant Danish tribes into a single kingdom. The maximum data rate and the maximum range of various versions of Bluetooth devices are shown in Table 3.11.

The Bluetooth module (HC-05) shown in Figure 3.73 has an antenna, a CSR Bluetooth controller, 8 MB flash memory, and a 26 MHz crystal oscillator.

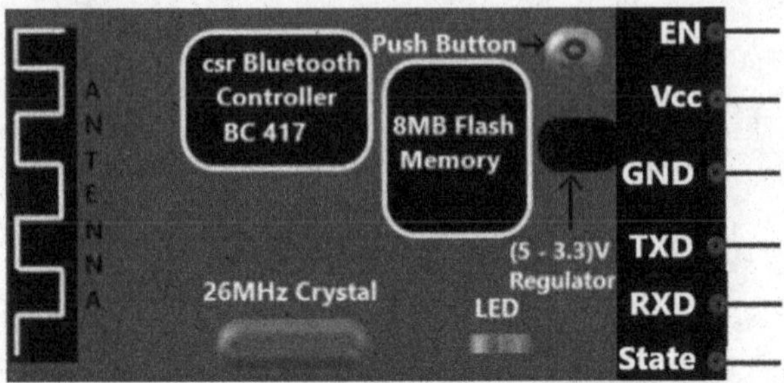

FIGURE 3.73 The front side of Bluetooth module (HC-05).

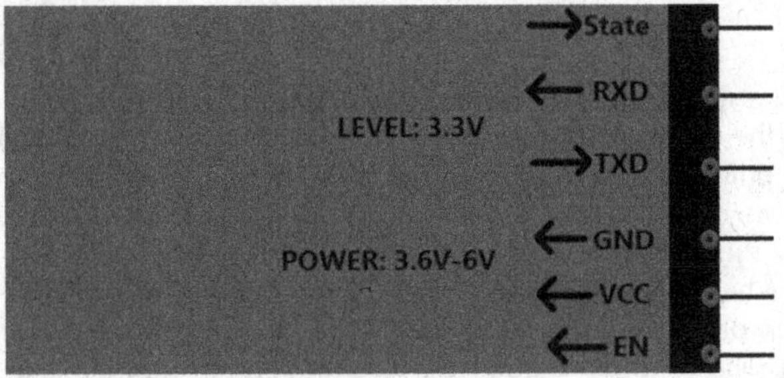

FIGURE 3.74 The back side of Bluetooth module (HC-05).

TABLE 3.11
The Bluetooth Versions Parameters

Version	Maximum Data Rate (Mbit/s)	Maximum Range (m)
3.0	25	10
4.0	25	60
5.0	50	240

An on-board 5–3.3 V regulator is there on the Bluetooth module (HC-05). The Bluetooth module (HC-05) operates at 3.3 V, but if we have connected the VCC pin of the module to the 5 V pin of Arduino, this on-board 5–3.3 V regulator converts 5 V into 3.3 V. There is a LED whose blinking sequence indicates whether the module is ready for data communication.

3.19.1 Pin Description and Other Details of Bluetooth Module (HC-05)

The Bluetooth module (HC-05) has six pins, and the description of these pins is as follows:

VCC: The VCC pin supplies the power to generate the ultrasonic pulses. It is to be connected to 5 or 3.3 V.

GND: The GND pin is connected to the ground.

TXD: The TXD pin of the Bluetooth module is a serial data transmission pin. This pin should be connected to the RX pin (Pin 0) of the Arduino board. Through the TXD pin, the Bluetooth module (HC-05) sends data serially to the Arduino board.

RXD: The RXD pin of the Bluetooth module is a serial data reception pin. This pin should be connected to the TX pin (Pin 1) of the Arduino board. Through the RXD pin, the Bluetooth module (HC-05) receives data serially from the Arduino board.

** Refer to Section 4.20 of Chapter 4 for programming and interfacing of the Bluetooth module with Arduino UNO.*

3.20 GSM MODULE (SIM900A)

The GSM module (SIM900A) is used for communication with remote locations provided the mobile tower is in range. It uses the same technology as used by the mobile communication network. SIM 900A is an example of a GSM module that supports standard AT commands. It is a tri-band GSM engine that works on frequencies 900, 1,800, and 1,900 MHz. A SIM 900A GSM module is shown in Figure 3.75.

The pin description and other details of the SIM900A GSM module are as follows:

SIM Socket: The SIM is inserted in this slot.

Adapter Socket: To power on the GSM module, we have to use 12 V, 2 A adapter. Please note the current rating of the adapter; if it is less than 2 A, then the GSM module may not function properly because when the GSM

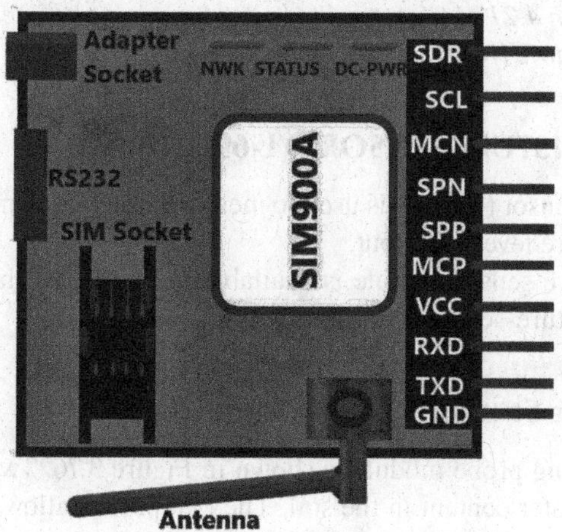

FIGURE 3.75 A SIM 900A GSM module.

module is communicating, it draws a lot of currents and this current varies from network to network so for the safer side, use adapter preferably 2 A.

SIM900A Modem: This modem works on AT commands. These AT commands are fed through the serial communication mode. For serial communication, there is an RS232 port available on the GSM module. But many microcontrollers don't have RS232 standards so data can be transferred serially to or from GSM module using TTL pins.

The Arduino board and SIM900A GSM module both follow a 5 V TTL level.

TXD Pin: The TXD pin of the GSM module should be connected to the RX pin (Pin 0) of the Arduino board.

RXD Pin: The RXD pin of the GSM module should be connected to the TX pin (Pin 1) of the Arduino board.

GND Pin: The GND pin of the GSM module should be connected to the Arduino board's GND pin.

VCC: To power on the GSM module, this pin should be connected to 7–12 V. This is an alternate way to power on the GSM module.

DC-PWR LED: The DC-PWR is power LED. It is on when the GSM module is turned on.

STATUS LED: The STATUS LED should also on when the GSM module is turned on. If the status LED is toggling, the current is not sufficient for the GSM module.

NWK LED: The NWK LED is network LED. It keeps on toggling fast when the GSM module searches for the mobile network. When the toggling rate of network LED is low, this means that the SIM is successfully registered.

The GSM module takes 10–15 seconds for registering itself. So send the AT commands once the GSM module is registered.

* *Refer to Section 4.21 of Chapter 4 for programming and interfacing of the GSM module with Arduino UNO.*

3.21 SOIL MOISTURE SENSOR (YL-69)

The soil moisture sensor (YL-69) is used to measure water content inside the soil and gives us the moisture level as output.

The soil moisture sensor module is available in two parts: the moisture-sensing probe and the moisture-sensing control board.

3.21.1 Moisture-Sensing Probe Module

The moisture-sensing probe module is shown in Figure 3.76. Two moisture-sensing probes sense the water content in the soil. The two probes allow the current to pass through the soil. The soil with more water will conduct more electricity, and the dry soil will conduct less. Thus, soil with more water will conduct more electricity – this

FIGURE 3.76 Moisture-sensing probe module.

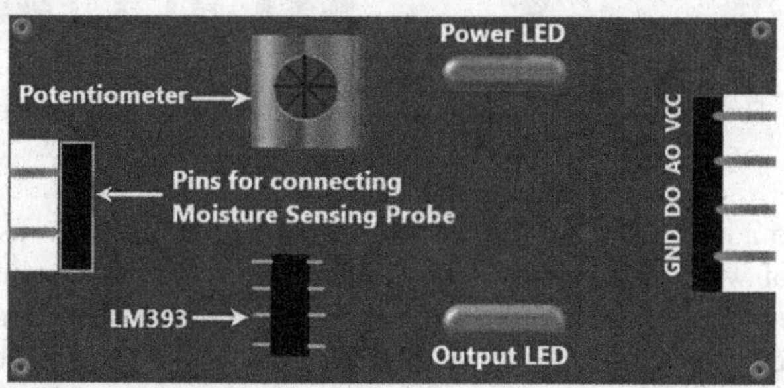

FIGURE 3.77 Moisture-sensing control board.

means its resistance is less; and the dry soil will conduct less electricity – this means its resistance is more.

3.21.2 Moisture-Sensing Control Module

The moisture-sensing control board is shown in Figure 3.77. The control board has four pins on one side and two pins on another side. In this module, we have to connect the VCC marked pin to 5 V, and GND marked pin is to be connected to the ground. The moisture sensor module comes with two different outputs, i.e., digital output (DO) and analog output (AO). The analog output voltage provided by the sensor changes is proportional to water content in the soil. The digital output will be high when the water content in the soil exceeds the sensitivity level set by the potentiometer.

There is a power LED which turns on when the module is appropriately power on by connecting VCC and GND pins at specified voltage levels. The potentiometer

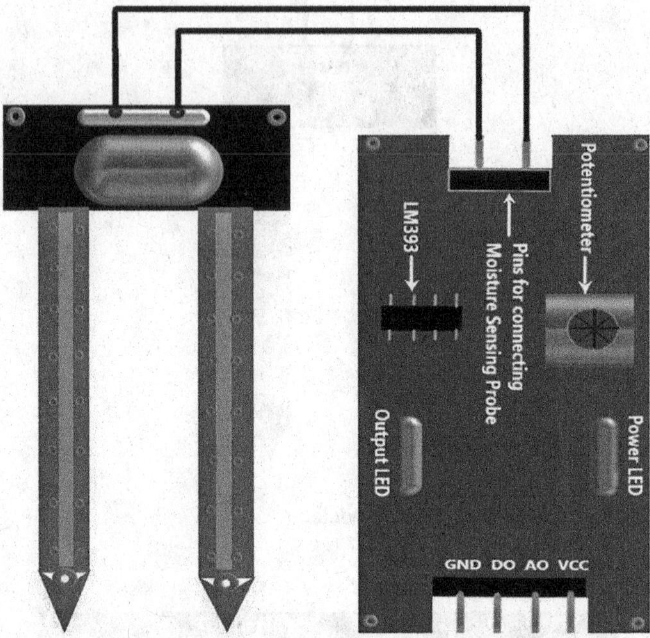

FIGURE 3.78 The soil moisture probe module and soil moisture-sensing control board.

can be used to adjust the sensitivity of the sensor module. We can use it to adjust the intensity of water it detects in soil. Suppose we rotate the potentiometer to a clockwise direction. In that case, the sensor becomes more sensitive, and if we rotate the potentiometer to an anti-clockwise direction, the sensor becomes less sensitive. When the water is detected, the output LED turns on. The onboard LM393 comparator IC on the control board is used for signal conditioning.

The moisture-sensing probe module and moisture-sensing control board are connected, as shown in Figure 3.78. The technical specifications of the soil moisture sensor (YL-69) are as follows:

i. Operation voltage: 3.3–5 V
ii. Output voltage: 0–4.2 V
iii. Input current: 35 mA
iv. Output: both analog output (AO) and digital output (DO).

3.21.3 Working of Moisture Sensor

The two pins of the control board available on another side of the board are to be connected to the two pins of the moisture-sensing probe module. These pins can be connected to the moisture-sensing control board in any direction. The probes of the module are inserted in the soil whose moisture is to be sensed. Depending upon the soil's water content, a proportional analog voltage or digital value will be generated from the analog output (AO) pin or digital output (DO) pin of the control board, respectively.

* *Refer to Section 4.22 of Chapter 4 for programming and interfacing of the moisture sensor with Arduino UNO.*

Check Yourself

1. The LED becomes forward bias when anode terminal is connected to the positive end and cathode terminal is connected to the negative end of the supply. (True/False)
2. If the anode of a LED is connected to 5 V and the cathode is connected to 2 V, then the LED will become forward bias. (True/False)
3. A push-button switch becomes short circuit when we press the switch. (True/False)
4. How many pins are there in a seven-segment package?
5. In order to turn on a segment of a common anode seven-segment display, what binary value is to be sent to segment?
6. What control word is needed to display "8" in a common anode-type seven-segment display?
7. What should be the logic level of Resister Select (RS) pin of LCD module in order to access command register?
8. What should be the logic level of RD/WR′ pin of LCD if we wish to write data word in data register of LCD module?
9. What is the significance of "B" if a potentiometer is labeled as "B10K"?
10. In Figure 3.79, if R1 = 5 KΩ and R2 = 5 KΩ, then what is the value of Vout?
11. What is the value of step size in in-built ADC of Arduino board?
12. The pin number 10 is the PWM pin of Arduino UNO board. (True/False)
13. Calculate the "value" required for generating a pulse width-modulated wave of 65% duty cycle at PWM pin number 9 of Arduino UNO board in the given function: `analogWrite(9,value);`
14. How much analog voltage will be generated at PWM Pin 9 of Arduino UNO board of Question 13 if a pulse of 65% duty cycle is generated at Pin 9?
15. The capacitance of a parallel plate capacitor increases when the distance between two plates reduces. (True/False)

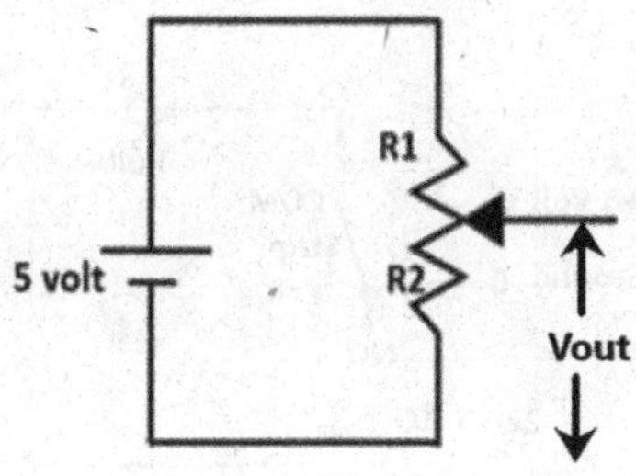

FIGURE 3.79 Circuit diagram of a potentiometer for Question 10.

16. The capacitive reactance decreases with the increase in capacitance of a capacitor. (True/False)
17. If in an application we want to rotate a motor in both clockwise and anti-clockwise directions, then how many maximum number of motor we can interface with L293D motor driver?
18. If in a 5 V DC motor we connect 3 V to both the terminals, then motor will rotate or not. If it rotates, then what is the direction of rotation?
19. Whether the bulb is on or off if it is connected to a relay as shown in Figure 3.80.
20. Whether the bulb is on or off if it is connected to a relay as shown in Figure 3.81.
21. The resistance of a LDR increases when more light falls on it. (True/False)
22. A rectangular pulse train has an on period of 150 μs and an off period of 1.5 ms. Calculate the percentage duty cycle of the signal.
23. Suppose a LED is interfaced to a port line in active low mode (glow when 0 is output) with a voltage source of 3.5 V. For passing 6 mA current, the current limiting resistance to be used in series with the LED must have the value ________ Ω, assuming a voltage drop of 1.4 V across the LED in forward-biased mode.
24. Which of the following statements is/are true for light-dependent resistor (LDR)?
    a. The resistance value decreases when light falls on it.
    b. The variation of resistance with illumination is non-linear.
    c. The resistance value increases when light falls on it.
    d. The variation of resistance with illumination is linear.

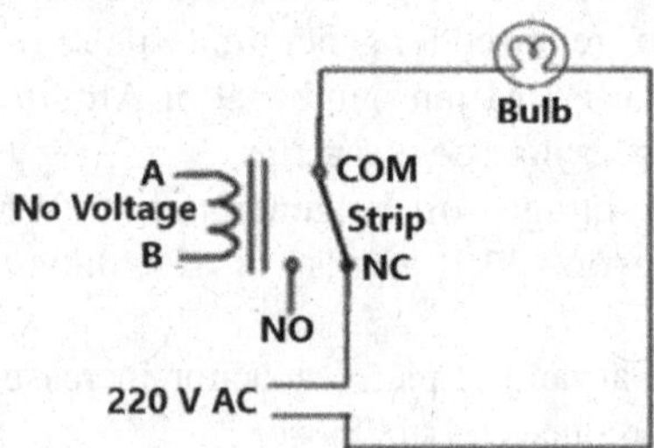

FIGURE 3.80 Interfacing diagram of a a bulb with relay for Question 19.

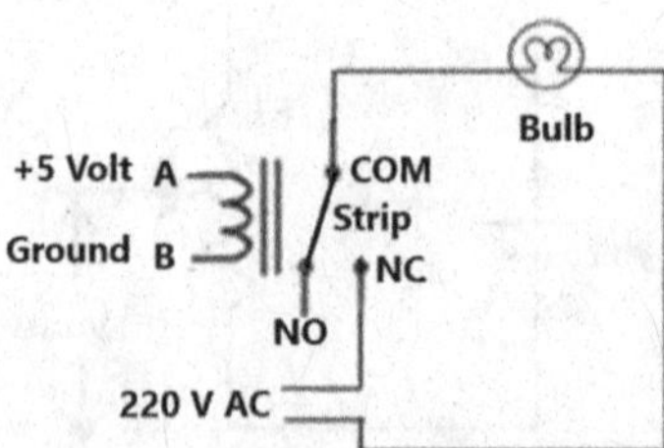

FIGURE 3.81 Interfacing diagram of a bulb with relay for Question 20.

25. An LDR interface is designed as follows. One terminal of the LDR is connected to 10 V, and its other terminal is connected to a resistance of 15 KΩ; the other terminal of the resistance is connected to ground. The voltage output Vout is taken from the junction point where the LDR is connected to the resistance. In the absence and in the presence of light, the LDR resistance value is 200 and 4 KΩ, respectively. What will be the voltage output at Vout in the absence and in the presence of light?
    a. 0.698 and 7.895 V
    b. 0.725 and 8.125 V
    c. 0.823 and 8.355 V
    d. None of these.
26. Which of the following is/are true for the LM35 temperature sensor?
    a. The variation in output voltage with temperature is linear.
    b. The variation in output voltage with temperature is non-linear.
    c. We require an A/D converter to interface it with the Arduino UNO microcontroller board.
    d. We do not require an A/D converter to interface it with the Arduino UNO microcontroller board.
27. What do the normally closed (NC) and normally open (NO) outputs of a relay module indicate?
    a. When the relay is not activated (normal state), the NC and the COM (pole) terminals of the output are connected.
    b. When the relay is activated (triggered state), the NC and the COM (pole) terminals of the output are connected.
    c. When the relay is not activated (normal state), the NO and the COM (pole) terminals of the output are connected.
    d. When the relay is activated (triggered state), the NO and the COM (pole) terminals of the output are connected.
28. Which of the following is/are true for the MQ2 gas sensor module?
    a. It is most suitable for detecting the presence of gases like methane, butane, smoke, etc.
    b. An analog output pin produces a voltage proportional to the intensity of the gas.
    c. A digital output pin produces a voltage proportional to the intensity of the gas.
    d. It requires a temperature sensor for its operation.
29. If we touch the touch pad of the sensor, then digital output pin of the module generates 0 V (Logic 0). (True/False)
30. When there is water drop on rain board, its resistance increases. (True/False)
31. The signals below 20 KHz are referred as ultrasonic signals.
32. The TXD and RXD pins of Bluetooth module should be connected to the pin numbers 1 and 0 of Arduino board, respectively. (True/False)
33. The TXD and RXD pins of GSM module should be connected to the pin numbers 0 and 1 of Arduino board, respectively. (True/False)

4 Interfacing and Programming with Arduino

LEARNING OUTCOMES

After completing this chapter, learners will be able to:

1. Demonstrate the interfacing and programming of light-emitting diode (LED), seven-segment display, and liquid crystal display (LCD) with Arduino UNO board.
2. Demonstrate the interfacing and programming of switch, keypad matrix, and potentiometer with Arduino UNO board.
3. Demonstrate the pulse width modulation (PWM) technique in controlling the operation of hardware.
4. Demonstrate the interfacing and programming of DC motor and motor driver board (L293D) with Arduino UNO board.
5. Demonstrate the interfacing and programming of relay board with Arduino UNO board to control high-voltage devices.
6. Demonstrate the interfacing and programming of temperature sensor (LM35), humidity and temperature sensor (DHT11), light-dependent register, touch sensor, smoke detector (MQ2), rain detector (FC-07), ultrasonic sensor (HC-SR04), and moisture sensor (YL-69) with Arduino UNO board.
7. Demonstrate the interfacing and programming of Bluetooth module (HC-05) and GSM module (SIM 900A) with Arduino UNO board.

4.1 LED INTERFACING AND PROGRAMMING

In this section, we shall discuss some programs to toggle the LED. These programs will enable readers to learn about the program flow. The readers will also understand how to interface a LED with the Arduino UNO board. The working principle of LED is explained in Section 3.1 of Chapter 3.

Program 4.1

Write a program for the Arduino UNO board to toggle an on-board LED connected at Pin 13.

Solution

The on-board LED connected to Pin 13 of the Arduino UNO board is shown in Figure 4.1. The Arduino UNO board program to toggle an LED connected at Pin 13 is shown in Figure 4.2.

Description of the Program:

By using statement (1), we wish to give the name LED to Pin 13. Here, LED is declared as a variable of integer type, and its assigned value is 13.

Inside `setup()` in the statement (2), the `pinMode` function declares Pin 13 as the output pin. Since Pin 13 is assigned to variable LED in the statement (1), in the statement (2) we have used the LED name of Pin 13 for initializing it as an output pin.

In this program, we wish to toggle an on-board LED connected at Pin 13. The LED is connected such that a high (5 V) at Pin 13 will make LED on and a low (0 V) at Pin 13 will make LED off.

FIGURE 4.1 The on-board LED connected to Pin 13 of the Arduino UNO board.

int LED=13;	*statement (1)*
void setup()	
{	
pinMode(LED,OUTPUT);	*statement (2)*
}	
void loop()	
{	
digitalWrite(LED,HIGH);	*statement (3)*
delay(1000);	*statement (4)*
digitalWrite(LED,LOW);	*statement (5)*
delay(2000);	*statement (6)*
}	

FIGURE 4.2 An Arduino UNO program to toggle an on-board LED connected at Pin 13 of Arduino UNO board.

Inside `loop()`, the `digitalWrite` function in the statement (3) is used to set Pin 13 to high. Due to statement (3), Pin 13 will set at 5 V. Since the anode and cathode of on-board LED are connected to Pin 13 and ground, respectively, after the execution of the `digitalWrite(LED, HIGH)` statement, the LED will be on.

The statement (4) is `delay(1000)`. It will generate a delay of 1,000 ms, i.e., 1 second. The `digitalWrite` function in the statement (5) is used to set Pin 13 to low (0 V). Due to statement (5), Pin 13 will set at 0 V. Since the anode and cathode of on-board LED are connected to Pin 13 and ground, respectively, after the execution of the `digitalWrite(LED, LOW)` statement, the LED will off. The statement (6) is `delay(2000)`. It will generate a delay of 2000 ms, i.e., 2 seconds. Due to statement (3), the LED is on, statement (4) will generate a delay of 1 second, statement (5) will make LED off, and statement (6) will generate a delay of 2 seconds. The statements (3), (4), (5), and (6) will make LED on for 1 second and off for 2 seconds. Since statement (6) is the last statement of the `loop()`, after statement (6) again statements (3), (4), (5), and (6) will be executed, and this process will continue. Due to the execution of statements as explained above, the LED will continue toggling until the Arduino board is getting the power.

Program 4.2

Write a program to toggle an externally connected LED at Pin 13 of Arduino UNO board.

Solution

The program for Arduino UNO board to toggle an externally connected LED at Pin 13 is shown in Figure 4.2. The circuit diagram of interfacing a LED with Pin 13 of the Arduino UNO board is shown in Figure 4.3. The anode of LED is connected to Pin 13 through 250 Ω resistor, and the cathode terminal is connected to the GND (ground) of the Arduino UNO board. The description of the circuit diagram is given in Section 3.1.3 of Chapter 3. The description of program is same as given in Program 4.1.

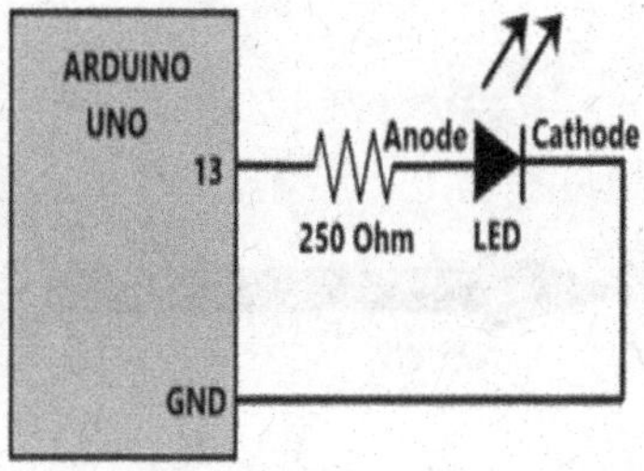

FIGURE 4.3 Circuit diagram for interfacing a LED with Pin 13 of Arduino UNO board.

Program 4.3

Write a program and develop an interfacing circuit with the Arduino UNO board for implementing traffic light logic.

Solution

The traffic light logic, circuit diagram to implement traffic light logic and Arduino program of traffic logic is shown in Figure 4.4, 4.5 and 4.6 respectively.

Traffic Light Logic: The three LEDs, their on-off sequences, and various traffic light logic states are shown in Figure 4.4. There are three LEDs, namely, red, orange, and green colors. The traffic light control logic can be described as follows:

State 1: In State 1, the red LED is on, while orange and green LEDs are off. The system will remain in this state for 10 seconds.

State 2: In State 2, the red LED is continued to be on, the orange LED is on, and the green LED is continued to be off. The system will remain in this state for 5 seconds. At the end of State 2, the red LED is on for 15 seconds, the orange LED is on for 5 seconds, and the green LED is off for 15 seconds.

State 3: In State 3, the red LED is off, the orange LED is off, and the green LED is on. The system will remain in this state for 10 seconds.

State 4: In State 4, the red LED is continued to be off, the orange LED is on, and the green LED is continued to be on. The system will remain in this state for 5 seconds. At the end of State 3 and State 4, the red LED is off for 15 seconds, the orange LED is on for 5 seconds, and the green LED is on for 15 seconds.

Circuit Diagram: The circuit diagram of an interfacing circuit with the Arduino UNO board to implement traffic light logic is shown in Figure 4.5. The red LED is connected to Pin 11, the orange LED is connected to Pin 12,

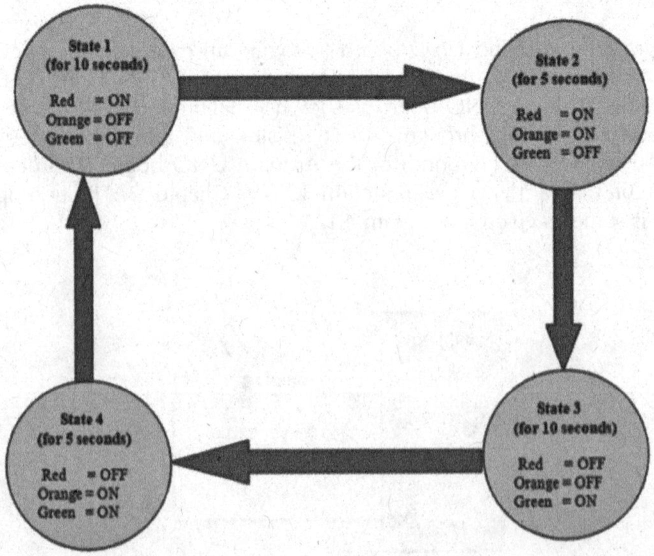

FIGURE 4.4 The traffic light logic.

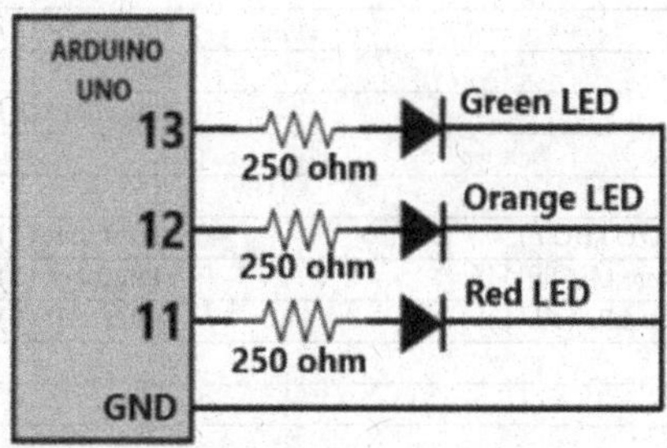

FIGURE 4.5 Circuit diagram of the implementation of traffic light logic.

and the green LED is connected to Pin 13 of the Arduino UNO board. The anode terminal of LEDs is connected to the Arduino board through 250 Ω resistors to save it from burning. The cathode terminal of LEDs is connected to the GND (ground) pin of the Arduino board.

Description of the Program:

By using the statements (1), (2), and (3), we give name red, orange, and green to Pins 11, 12, and 13 of the Arduino UNO board.

Inside `setup()` in the statements (4), (5), and (6), `pinMode` function is used to declare Pins 11, 12, and 13 as output pins.

The anode terminal of three LEDs is connected to Pins 11, 12, and 13 of the Arduino board through 250 Ω resistor, and the cathode terminal of three LEDs is connected to the GND (ground) as shown in Figure 4.5. A high (5 V) at Pins 11, 12, and 13 will make red, orange, and green LEDs on, respectively, and a low (0 V) at Pins 11, 12, and 13 will make red, orange, and green LEDs off, respectively.

Inside `loop()`, the `digitalWrite` function in the statements (7), (8), and (9) is used to set Pin 11 and to reset Pins 12 and 13. The statements (7), (8), and (9) will implement State 1 of Figure 4.4. The delay of 10 seconds of State 1 will be implemented by statement (10).

The statements (11), (12), and (13) are used to set Pins 11 and 12 and to reset Pin 13. The statements (11), (12), and (13) will implement State 2 of Figure 4.4. The delay of 5 seconds of State 2 will be implemented by statement (14).

The `digitalWrite` function in the statements (15), (16), and (17) is used to reset Pins 11 and 12 and to set Pin 13. The statements (15), (16), and (17) will implement State 3 of Figure 4.4. The delay of 10 seconds of State 3 will be implemented by statement (18).

The statements (19), (20), and (21) are used to reset Pin 11 and to set Pins 12 and 13. The statements (19), (20), and (21) will implement State 4 of Figure 4.4. The delay of 5 seconds of State 4 will be implemented by statement (22). The execution of statements as explained above will toggle the three LEDs

int red=11;	statement (1)
int orange=12;	statement (2)
int green=13;	statement (3)
void setup()	
{	
pinMode(red,OUTPUT);	statement (4)
pinMode(orange,OUTPUT);	statement (5)
pinMode(green,OUTPUT);	statement (6)
}	
void loop()	
{	
digitalWrite(red,HIGH);	statement (7)
digitalWrite(orange,LOW);	statement (8)
digitalWrite(green,LOW);	statement (9)
delay(10000);	statement (10)
digitalWrite(red,HIGH);	statement (11)
digitalWrite(orange,HIGH);	statement (12)
digitalWrite(green,LOW);	statement (13)
delay(5000);	statement (14)
digitalWrite(red,LOW);	statement (15)
digitalWrite(orange,LOW);	statement (16)
digitalWrite(green,HIGH);	statement (17)
delay(10000);	statement (18)
digitalWrite(red,LOW);	statement (19)
digitalWrite(orange,HIGH);	statement (20)
digitalWrite(green,HIGH);	statement (21)
delay(5000);	statement (22)
}	

FIGURE 4.6 An Arduino UNO program to implement traffic light logic for the circuit diagram shown in Figure 4.5.

connected to Pins 11, 12, and 13 as per the traffic light logic shown in Figure 4.4 till the Arduino board is getting the power.

The statements (7), (8), (9), and (10) are used to implement State 1. The statements (11), (12), (13), and (14) are used to implement State 2. The statements (15), (16), (17), and (18) are used to implement State 3. The statements (19), (20), (21), and (22) are used to implement State 4.

Modification in the program: As per the traffic light logic as shown in Figure 4.4, there is no change in the state of red and green LEDs of State 1 and State 2; therefore, we can remove statements (11) and (13) of the program shown in Figure 4.6. There is no change in the state of red and green LEDs of State 3 and State 4; therefore, we can remove statements (19) and (21) of the program shown in Figure 4.6. The modified program is shown in Figure 4.7. As a system design engineer, we must develop a program in minimum statements because a smaller code requires less memory for dumping it.

4.2 DISPLAY IN SERIAL MONITOR

In this section, we shall discuss some programs to understand the working principle of various functions of "Serial" library to display the contents on the serial monitor. The

int red=11;	statement (1)
int orange=12;	statement (2)
int green=13;	statement (3)
void setup()	
{	
pinMode(red,OUTPUT);	statement (4)
pinMode(orange,OUTPUT);	statement (5)
pinMode(green,OUTPUT);	statement (6)
}	
void loop()	
{	
digitalWrite(red,HIGH);	statement (7)
digitalWrite(orange,LOW);	statement (8)
digitalWrite(green,LOW);	statement (9)
delay(10000);	statement (10)
digitalWrite(orange,HIGH);	statement (11)
delay(5000);	statement (12)
digitalWrite(red,LOW);	statement (13)
digitalWrite(orange,LOW);	statement (14)
digitalWrite(green,HIGH);	statement (15)
delay(10000);	statement (16)
digitalWrite(orange,HIGH);	statement (17)
delay(5000);	statement (18)
}	

FIGURE 4.7 A modified Arduino UNO program to implement four states of traffic light logic for the circuit diagram shown in Figure 4.5.

display on the serial monitor of Arduino IDE is initiated by using `Serial.print` and `Serial.println` functions. The programs in this section will make readers to understand how to use the `Serial.print` and `Serial.println` functions wisely.

Program 4.4

Write a program for Arduino UNO board to display on the serial monitor using `Serial.print` function.

Solution

The program using `Serial.print` function to display on the serial monitor is shown in Figure 4.8.

Description of the Program:

The *`Serial.begin(9600)`* function of the statement (1) will initialize the serial communication between the Arduino board and the computer. The data transmission rate is set to 9,600 baud, and Pins 0 and 1 will be used for serial data reception and transmission, respectively.

The *`Serial.print("MY  SWEET  HOME")`* function of the statement (2) will print "MY SWEET HOME" on the serial monitor, and the cursor will remain in the same line after printing.

void setup()	
{	
Serial.begin(9600);	statement (1)
}	
void loop()	
{	
Serial.print("MY SWEET HOME");	statement (2)
delay(1000);	statement (3)
}	

FIGURE 4.8 Program to display on the serial monitor by using "Serial.print" function.

The `delay(1000)` function of the statement (3) will generate a delay of 1,000 ms (1 second), and microcontroller will not execute another statement during the delay period.

Since statement (3), i.e., `delay(1000),` is the last instruction of the `void loop()`, the first statement of the loop, i.e., `Serial.print("MY SWEET HOME"),` will again be executed. In this way, the `void loop()` will run continuously till the Arduino UNO is getting power.

The screenshot of serial monitor of Program 4.4 displaying MY SWEET HOME continuously in the same line on the serial monitor is shown in Figure 4.9.

Modification in program: The `Serial.begin(9600)` function of the statement (1) will initialize the serial communication between the Arduino board and the computer at 9,600 Baud. The `Serial.print("MY")` function of the statement (2) will print "MY" on the serial monitor, and the cursor will remain in the same line after printing.

The `Serial.print(" SWEET")` function of the statement (3) will print " SWEET" on the serial monitor in continuation with "MY", and after the execution of statement (3), "MY SWEET" will be displayed on the serial monitor. The cursor will remain in the same line after printing. Readers observe the space between MY and SWEET. The space is due to the space SWEET in the statement (3).

The `Serial.print(" HOME")` function of the statement (4) will print " HOME" on the serial monitor in continuation with "MY SWEET", and after the execution of statement (4), "MY SWEET HOME" will be displayed on the serial monitor. The cursor will remain in the same line after printing. Readers observe the space between MY SWEET and HOME. The space is due to the space HOME in the statement (4).

Due to the `delay(1000)` function of the statement (5), a delay of 1 second will be generated. Since the `delay(1000)` is the last instruction of the `void loop()`, all the statements inside the loop will be executed one after another in sequence until the Arduino UNO is getting power. The programs shown in Figures 4.8 and 4.10 generate the same result as shown in Figure 4.9.

Program 4.5

Write a program for Arduino UNO board to display on the serial monitor using `Serial.println` function.

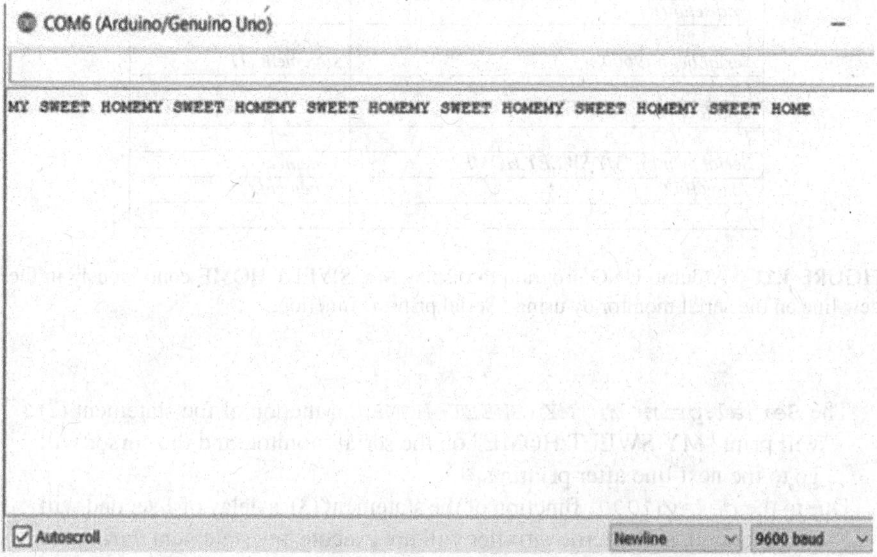

COM6 (Arduino/Genuino Uno)

MY SWEET HOMEMY SWEET HOMEMY SWEET HOMEMY SWEET HOMEMY SWEET HOMEMY SWEET HOME

Autoscroll Newline 9600 baud

FIGURE 4.9 The screenshot of serial monitor of program shown in Figure 4.8 displaying MY SWEET HOME continuously in the same line on the serial monitor.

void setup()	
{	
Serial.begin(9600);	*statement (1)*
}	
void loop()	
{	
Serial.print("MY");	*statement (2)*
Serial.print(" SWEET");	*statement (3)*
Serial.print(" HOME");	*statement (4)*
delay(1000);	*statement (5)*
}	

FIGURE 4.10 A modified Arduino UNO program to display MY SWEET HOME continuously in the same line on the serial monitor by using "Serial.print" function.

Solution

The Arduino UNO program to display MY SWEET HOME continuously in the new line on the serial monitor by using "Serial.println" function is shown in Figure 4.11.

Description of the Program:

The `Serial.begin(9600)` function of the statement (1) will initialize the serial communication between the Arduino board and the computer at 9,600 Baud.

void setup()	
{	
Serial.begin(9600);	*statement (1)*
}	
void loop()	
{	
Serial.println("MY SWEET HOME");	*statement (2)*
delay(1000);	*statement (3)*
}	

FIGURE 4.11 Arduino UNO program to display MY SWEET HOME continuously in the new line on the serial monitor by using "Serial.println" function.

The `Serial.println("MY SWEET HOME")` function of the statement (2) will print "MY SWEET HOME" on the serial monitor, and the cursor will go to the next line after printing.

Due to the `delay(1000)` function of the statement (3), a delay of 1 second will be generated, and microcontroller will not execute any statement during the delay period.

Since the `delay(1000)` is the last instruction of the void loop(), the loop's first statement, i.e., `Serial.println("MY SWEET HOME"),` will again be executed. In this way, the `void loop()` will run continuously till the Arduino UNO is getting power. The screenshot of the serial monitor is shown in Figure 4.12.

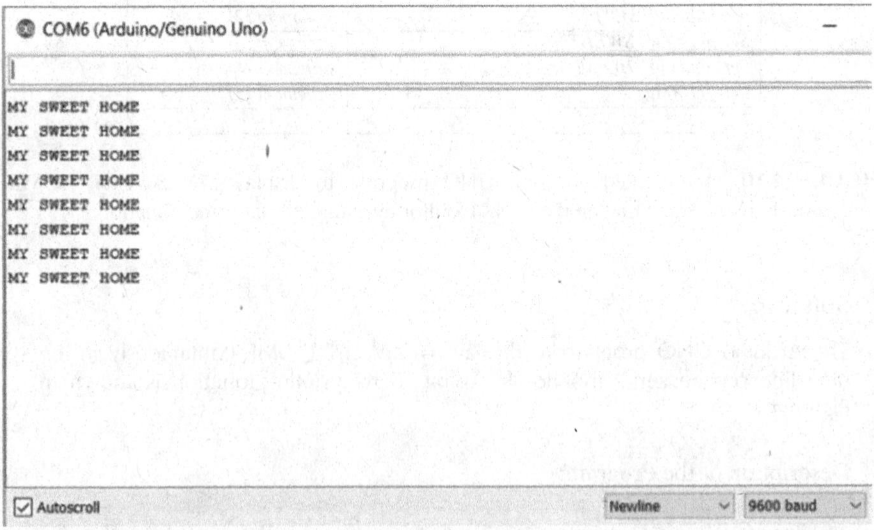

```
COM6 (Arduino/Genuino Uno)

MY SWEET HOME
MY SWEET HOME
MY SWEET HOME
MY SWEET HOME
MY SWEET HOME
MY SWEET HOME
MY SWEET HOME
MY SWEET HOME

Autoscroll                    Newline    9600 baud
```

FIGURE 4.12 The screenshot of serial monitor of program shown in Figure 4.11 displaying MY SWEET HOME continuously in the new line on the serial monitor.

Program 4.6

Write a program for Arduino UNO board to display numbers from 0 to 9 continuously on the serial monitor.

Solution

The Arduino UNO program to display numbers from 0 to 9 continuously on the serial monitor is shown in Figure 4.13.

Description of the Program:

The `Serial.begin(9600)` function of the statement (1) will initialize the serial communication between the Arduino board and the computer at 9,600 Baud.

In statement (2), `for (int i=0; i<=9; i++)` integer i is created with the initial value 0. Since the value of i ≤ 9 is true, the body of the loop, i.e., statements (3), (4), and (5), will be executed, and i will be incremented by one due to the presence of i++ in for loop.

The `Serial.print("i= ")` function of the statement (3) will print "i= " on the serial monitor and the cursor will remain on the same line after printing. The `Serial.println(i)` function in the statement (4) will print the value of i, which is 0 at this moment in continuation with "i=". So at the end of the statement (4), "i= 0" will be displayed on the serial monitor and the cursor will go to the next line after printing. Due to the `delay(1000)` function of the statement (5), a delay of 1 second will be generated, and microcontroller will not execute any statement during the delay period. After 1 second again, i ≤ 9 condition is evaluated. The value of i is 1, so i= 1 will be printed on the serial monitor in the next line. In this way, numbers will be printed one after the other with a gap of 1 second. Once i= 9 is printed, the value of i will be reset to 0 and printing from i= 0 continues again. The screenshot of the serial monitor is shown in Figure 4.14.

void setup()	
{	
Serial.begin(9600);	*statement (1)*
}	
void loop()	
{	
for(int i=0;i<=9;i++)	*statement (2)*
{	
Serial.print("i= ");	*statement (3)*
Serial.println(i);	*statement (4)*
delay(1000);	*statement (5)*
}	
}	

FIGURE 4.13 Arduino UNO program to display numbers from 0 to 9 continuously on the serial monitor.

```
COM6 (Arduino/Genuino Uno)

i= 0
i= 1
i= 2
i= 3
i= 4
i= 5
i= 6
i= 7
i= 8
i= 9
i= 0
i= 1
i= 2
i= 3

Autoscroll
```

FIGURE 4.14 The screenshot of serial monitor of program shown in Figure 4.13 displaying numbers from 0 to 9 continuously on the serial monitor.

Program 4.7

Write a program for Arduino UNO board to display numbers from 0 to 9 on the serial monitor only once.

Solution

The Arduino UNO program to display numbers from 0 to 9 only once on the serial monitor is shown in Figure 4.15.

Description of the Program:

The `Serial.begin(9600)` function of the statement (1) will initialize the serial communication between the Arduino board and the computer at 9,600 Baud.

Since the statements (2), (3), (4), and (5) are in `void setup()`, these statements will be executed only once.

Already we have discussed in the program shown in Figure 4.13 that the statements (2), (3), (4), and (5) will display i= 0 to i= 9. Therefore, program shown in Figure 4.15 will display numbers from 0 to 9 on the serial monitor once. The screenshot of serial monitor of program shown in Figure 4.15 displaying numbers from 0 to 9 only once on the serial monitor is shown in Figure 4.16.

void setup()	
{	
Serial.begin(9600);	*statement (1)*
for(int i=0;i<=9;i++)	*statement (2)*
{	
Serial.print("i= ");	*statement (3)*
Serial.println(i);	*statement (4)*
delay(1000);	*statement (5)*
}	
}	
void loop()	
{	
}	

FIGURE 4.15 Arduino UNO program to display numbers from 0 to 9 only once on the serial monitor.

```
COM6 (Arduino/Genuino Uno)

i= 0
i= 1
i= 2
i= 3
i= 4
i= 5
i= 6
i= 7
i= 8
i= 9

Autoscroll                                   Newline
```

FIGURE 4.16 The screenshot of serial monitor of program shown in Figure 4.15 displaying numbers from 0 to 9 only once on the serial monitor.

4.3 PUSH-BUTTON INTERFACING AND PROGRAMMING

This section shall discuss some programs to understand the interfacing and programming related to the push button. After going through this section, the reader will understand what value (Logic 0/Logic 1) will be fed to an Arduino board pin when a switch is pressed or not pressed. The working principle of push-button switch is explained in Section 3.2 of Chapter 3.

Program 4.8

A push-button switch is connected to Pin 2 of Arduino UNO board as shown in Figure 4.17. Write a program to read the position of the switch and display it on the serial monitor.

Solution

A push-button switch is connected to Pin 2 of Arduino UNO board. The interfacing diagram of a switch with Arduino UNO board is shown in Figure 4.17. The Terminal T2 of the push-button switch is connected to the GND (ground) pin of Arduino board, and the Terminal T1 is connected to the one terminal of 1 KΩ resistor. The other terminal of 1 KΩ resistor is connected to the 5 V pin of Arduino board. The junction of Terminal T1 of switch and 1 KΩ resistor is extended and connected to the pin number 2 of Arduino board. The working principle of the push-button switch is explained in Section 3.2 of Chapter 3. The program to read the switch's position and display it on the serial monitor is shown in Figure 4.18.

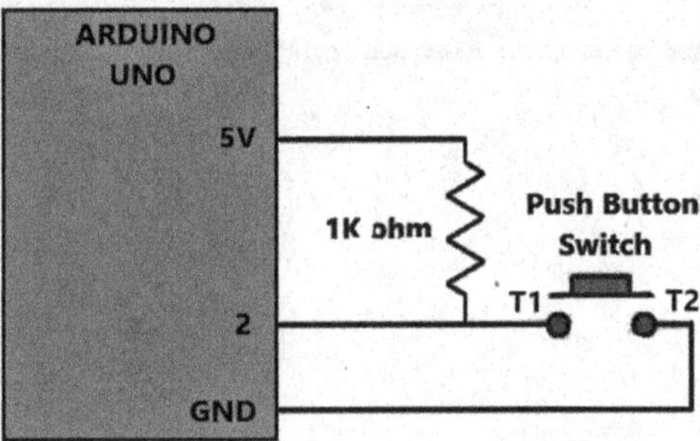

FIGURE 4.17 Interfacing of push-button switch for Program 4.8.

int pushButton=2;	*statement (1)*
void setup()	
{	
Serial.begin(9600);	*statement (2)*
pinMode(pushButton,INPUT);	*statement (3)*
}	
void loop()	
{	
int buttonState;	*statement (4)*
buttonState =digitalRead(pushButton);	*statement (5)*
Serial.println(buttonState);	*statement (6)*
delay(1000);	*statement (7)*
}	

FIGURE 4.18 An Arduino UNO program to display the position of push-button switch on the serial monitor for circuit diagram shown in Figure 4.17.

Description of the Program:

Using the statement (1), we give name "pushButton" to Pin 1 to which we have interfaced push-button switch.

The *Serial.begin(9600)* function of the statement (2) will initialize the serial communication between the Arduino board and the computer at 9,600 Baud.

The *pinMode* function in the statement (3) is used to declare the "pushButton" (Pin 2) as an input pin.

Using the statement (4), we have declared "buttonState" a variable of integer type.

The *buttonState =digitalRead(pushButton)* statement (5) is used to read the digital value of the variable "pushButton" (Pin 2) and assign this value to variable "buttonState". As per Figure 4.17 if the push-button is not pressed, then "1" (binary 1 or 5 V) will be assigned to variable "buttonState", and if the push-button is pressed, then "0" (binary 0 or 0 V) will be assigned to variable "buttonState".

The *Serial.println(buttonState)* function of the statement (6) will print the value of buttonState on the serial monitor, and the cursor will go to the next line after printing.

Due to the *delay(1000)* function of the statement (7), a delay of 1 second is generated.

The program shown in Figure 4.18 will print the push-button value (0 or 1) on the serial monitor after every second. The screenshot of the serial monitor is shown in Figure 4.19.

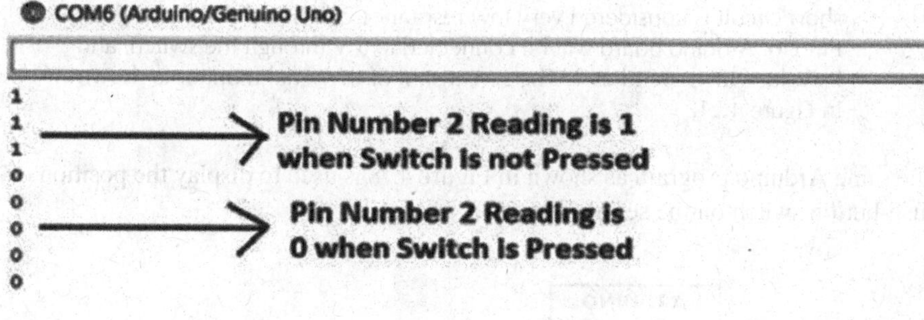

FIGURE 4.19 The screenshot of serial monitor to display push-button switch value on the serial monitor after every second of program shown in Figure 4.18 for circuit diagram shown in Figure 4.17.

Program 4.9

A push-button switch is connected to Pin 2 of Arduino UNO board as shown in Figure 4.20. Suppose the program shown in Figure 4.18 is used, then show the value of switch on the serial monitor, and explain it.

Solution

A push-button switch is connected to Pin 2 of Arduino UNO board. The interfacing diagram of a switch with Arduino UNO board is shown in Figure 4.20. The Terminal T2 of the push-button switch is connected to the 5 V pin of Arduino board, and the Terminal T1 is connected to the one terminal of 1 KΩ resistor. The other terminal of 1 KΩ resistor is connected to the GND (ground) pin of Arduino board. The junction of Terminal T1 of switch and 1 KΩ resistor is extended and connected to the pin number 2 of Arduino board. The working principle of push-button switch is explained in Section 3.2 of Chapter 3. The working principle of push-button switch and its interfacing with Arduino UNO can be explained by the following two cases:

Case 1: In Case 1 during an initial state when the pushing pad of switch is not pressed, a switch's open-circuit condition conducts no current through it. Pin 2 of Arduino always follows a low resistance path. Pin 2 is either connected to ground (GND) through resistor R1 or connected to 5 V through open-circuit switch. Since the resistance R1 (1 KΩ) is very small compared with open circuit (the open circuit is considered infinite resistance), Pin 2 will be connected to GND through resistor R1, and its value will be read as 0.

Case 2: In Case 2, when the pushing pad of switch is pressed, the short-circuit condition of switch conducts current through it. Pin 2 of Arduino always follows a low resistance path. Pin 2 is either connected to GND through resistor R1 or connected to 5 V through the short-circuit switch. Since the short circuit is considered very low resistance compared with R1 (1 KΩ), Pin 2 of Arduino board will be connected to 5 V through the switch, and its value will be read as 1. The screenshot of the serial monitor is shown in Figure 4.21.

The same Arduino program as shown in Figure 4.18 is used to display the position of push-button switch on the serial monitor.

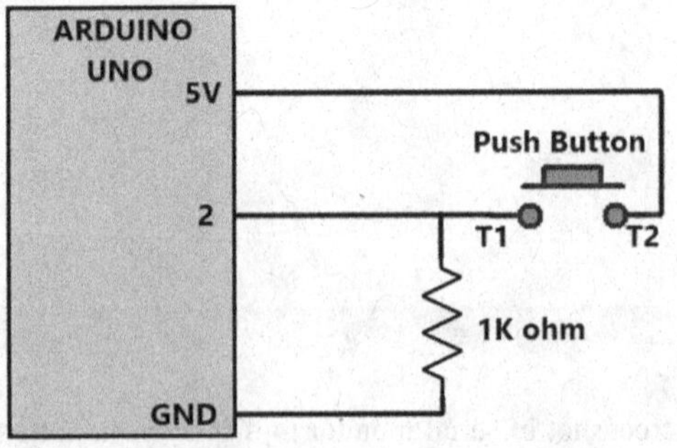

FIGURE 4.20 Interfacing of push-button switch for Program 4.9.

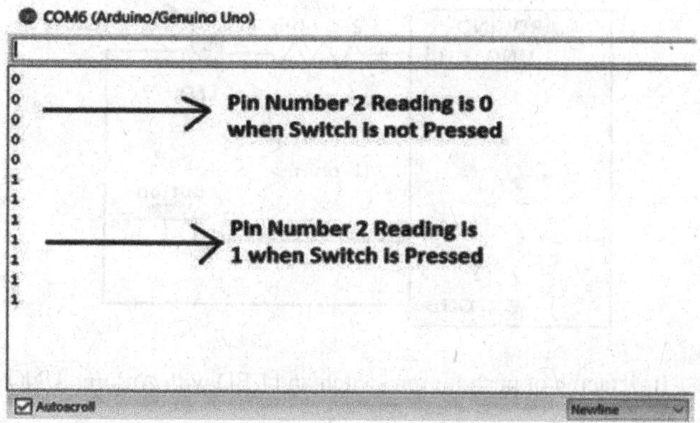

FIGURE 4.21 The screenshot of serial monitor of program shown in Figure 4.18 and circuit diagram shown in Figure 4.20 to display push-button switch value on the serial monitor after every second.

Program 4.10

Interface a push-button switch and LED with Arduino UNO board. Write a program to turn on and off LED when the switch is pushed and released, respectively.

Solution

The interfacing diagram of a push-button switch and LED with Arduino UNO board is shown in Figure 4.22. The Terminal T2 of the push-button switch is connected to the GND (ground) pin of Arduino board and the Terminal T1 is connected to the one terminal of 1 KΩ resistor. The other terminal of 1 KΩ resistor is connected to the 5 V pin of Arduino board. The junction of Terminal T1 of switch and 1 KΩ resistor is extended and connected to the pin number 2 of Arduino board. The anode of LED is connected to the pin number 13 of Arduino UNO board through a 250 Ω resistor, and the cathode is connected to the GND (ground) pin of Arduino board.

An Arduino UNO program is shown in Figure 4.23 to turn on and off LED when switch is pushed and released for the circuit diagram shown in Figure 4.22.

Description of the Program:

Using the statement (1), we give name "pushButton" to Pin 1 to which we have interfaced push-button switch.

Using the statement (2), we give name "LED" to Pin 13 to which we have interfaced a LED.

The `pinMode(pushButton, INPUT)` function in statement (3) is used to declare the "pushButton" (Pin 2) as input pin.

The `pinMode(LED, OUTPUT)` function in statement (4) is used to declare the "LED" (Pin 13) as an output pin.

The `int buttonState =digitalRead(pushButton)` statement will declare "buttonState" an integer type variable and read the digital value of

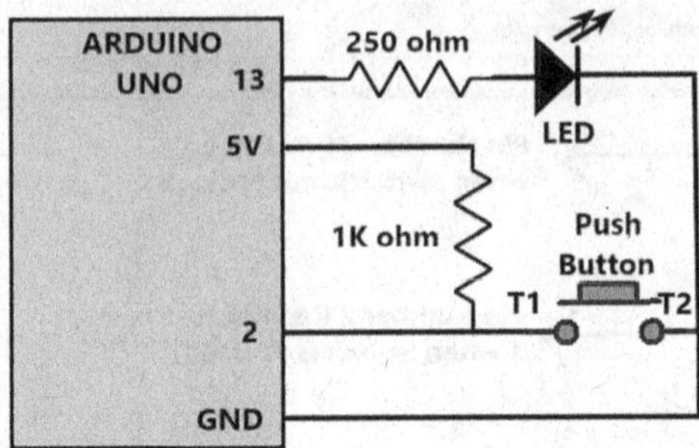

FIGURE 4.22 Interfacing of push-button switch and LED with Arduino UNO board.

int pushButton=2;	*statement (1)*
int LED=13;	*statement (2)*
void setup()	
{	
pinMode(pushButton,INPUT);	*statement (3)*
pinMode(LED,OUTPUT);	*statement (4)*
}	
void loop()	
{	
int buttonState =digitalRead(pushButton);	*statement (5)*
if (buttonState==0)	*statement (6)*
{	
digitalWrite(LED,HIGH);	*statement (7)*
}	
else	
{	
digitalWrite(LED,LOW);	*statement (8)*
}	
delay(500);	*statement (9)*
}	

FIGURE 4.23 An Arduino UNO program to turn on and off LED when switch is pushed and released for the circuit diagram shown in Figure 4.22.

pushButton (Pin 2) and assign it's value to buttonState. As per Figure 4.22 if the push button is not pressed, then "1" (5 V) will be assigned to variable "buttonState", and if the push-button is pressed, then "0" (0 V) will be assigned to variable "buttonState".

In statement (6) "if (buttonState==0)", the value of "buttonState" (i.e., Pin 2) is evaluated. If the push-button is pressed, then it's value will be 0, then statement (7) `digitalWrite(LED, HIGH)` will be executed, and LED will on; otherwise, statement (8) `digitalWrite(LED, LOW)` will be executed and LED will off.

Due to the `delay(500)` function of the statement (9), a delay of 1 ms will be generated.

Conclusion

a. If push-button is pressed, then Pin 2 of Arduino board will get 0, and this condition will execute statement (7) `digitalWrite(LED, HIGH)` and LED will on.
b. If the push button is not pressed, then Pin 2 of Arduino board will get 1, and this condition will execute statement (8) `digitalWrite(LED, LOW)` and LED will off.

Program 4.11

Interface a push-button switch and LED with Arduino UNO board as shown in Figure 4.24. Suppose the program shown in Figure 4.23 is used, then explain the working principle of the circuit.

Solution

The interfacing diagram of a push-button switch and LED with Arduino UNO board is shown in Figure 4.24. The Terminal T2 of the push button is connected to the 5 V pin of Arduino board, and the Terminal T1 is connected to the one terminal of 1 KΩ resistor. The other terminal of 1 KΩ resistor is connected to the GND (ground) pin of Arduino board. The junction of Terminal T1 of switch and 1 KΩ resistor is extended and connected to the pin number 2 of Arduino board. The anode of LED is connected to the pin number 13 of Arduino UNO board through a 250 Ω resistor, and the cathode is connected to the GND (ground) pin of Arduino board.

Description of the Program as shown in Figure 4.23:

Due to the change in the circuit diagram as shown in Figure 4.24, the program, as shown in Figure 4.23, will perform in the following way:

a. If push-button is not pressed, then Pin 2 of Arduino board will get 0, and this condition will execute statement (7) `digitalWrite(LED, HIGH)` and LED will on.
b. If the push-button is pressed, then Pin 2 of the Arduino board will get 1, and this condition will execute statement (8) `digitalWrite(LED, LOW)` and LED will off.

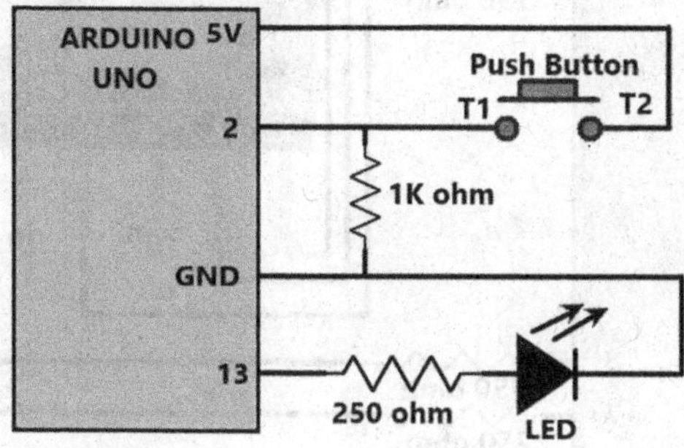

FIGURE 4.24 Interfacing of push-button switch and LED with Arduino UNO board.

4.4 SEVEN-SEGMENT DISPLAY INTERFACING AND PROGRAMMING

This section shall discuss the interfacing and programming related to seven-segment display. A seven-segment display is generally used to display numbers from 0 to 9. The working principle of seven-segment display (common cathode (CC) and common anode types) is explained in Section 3.3 of Chapter 3.

Example 4.1

Interface a CC-type seven-segment display with Arduino UNO board, and develop the control word to display numbers from 0 to 9.

Solution

The interfacing of a CC-type seven-segment display with Arduino UNO board is shown in Figure 4.25, and the pin-to-pin mapping of CC seven-segment display and Arduino UNO board is shown in Table 4.1. We know that each pin of Arduino UNO outputs 40 mA. In Section 3.1.3 of Chapter 3, we discussed that LED requires a current in the range of 15–20 mA depending upon the variety of LED. The output current from the Arduino board will burn out the LED if no protection is incorporated. It is shown in Figure 4.25 that a resistor of 250 Ω is connected between each segment and the pin of Arduino UNO to prevent the burning of the segment due to over-current. The segments a, b, c, d, e, f, and g of seven-segment display are connected to pin numbers 7, 8, 9, 10, 11, 13, and 12 of Arduino board through 250 Ω resistor. The com (common) pin of the seven-segment display is connected to GND (ground) pin of Arduino board.

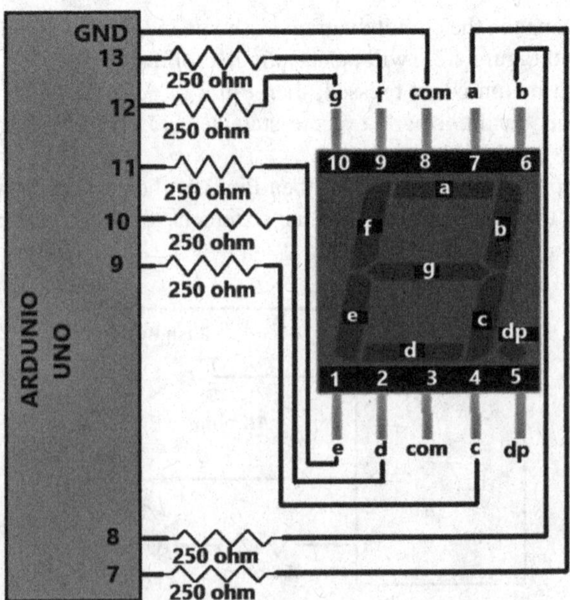

FIGURE 4.25 Interfacing of common cathode seven-segment display with Arduino UNO board.

TABLE 4.1
Pin-to-Pin Mapping of the Seven-Segment Display with Arduino UNO Board

CC Seven-Segment Display Pin	Arduino UNO Pin Number/Name
a	7
b	8
c	9
d	10
e	11
f	13
g	12
Common (com)	GND
dp	No connection

TABLE 4.2
Control Words to Display Numbers from 0 to 9 for CC Seven-Segment Display

Segments of Display								Number to Display
dp	g	f	e	d	c	b	a	
0	0	1	1	1	1	1	1	0
0	0	0	0	0	1	1	0	1
0	1	0	1	1	0	1	1	2
0	1	0	0	1	1	1	1	3
0	1	1	0	0	1	1	0	4
0	1	1	0	1	1	0	1	5
0	1	1	1	1	1	0	1	6
0	0	0	0	0	1	1	1	7
0	1	1	1	1	1	1	1	8
0	1	1	0	0	1	1	1	9

To turn on any segment in a CC-type seven-segment display, we have to send 5 V (binary 1) to the specific segment, and to turn off, we have to send 0 V (binary 0). The development of the control word is explained in Example 3.1 in Chapter 3. The control words to display numbers from 0 to 9 for CC seven-segment display are shown in Table 4.2.

Program 4.12

Interface a CC-type seven-segment display with Arduino UNO board, and write a program to display numbers from 0 to 9.

Solution

The interfacing of a CC-type seven-segment display with Arduino UNO board is shown in Figure 4.25, and the pin-to-pin mapping of CC seven-segment display and Arduino UNO board is shown in Table 4.1. The program to display numbers from 0 to 9 is shown in Figure 4.26.

Description of the Program:

Using the statements (1) to (7), Pins 7, 8, 9, 10, 11, 13, and 12 of Arduino UNO board are given the names as a, b, c, d, e, f, and g, respectively, of the seven-segment display.

Using the `pinMode` function from statements (8) to (14), Pins 7, 8, 9, 10, 11, 13, and 12 of Arduino UNO board are declared as output pins.

By using the `digitalWrite` function from statements (15) to (21), number 0 will be displayed, and `delay(1000)` function of the statement (22) will generate a delay of 1 second.

By using the `digitalWrite` function from statements (23) to (29), number 1 will be displayed, and `delay(1000)` function of the statement (30) will generate a delay of 1 second.

By using the `digitalWrite` function from statements (31) to (37), number 2 will be displayed, and `delay(1000)` function of the statement (38) will generate a delay of 1 second.

By using the `digitalWrite` function from statements (39) to (45), number 3 will be displayed, and `delay(1000)` function of the statement (46) will generate a delay of 1 second.

By using the `digitalWrite` function from statements (47) to (53), number 4 will be displayed, and `delay(1000)` function of the statement (54) will generate a delay of 1 second.

By using the `digitalWrite` function from statements (55) to (61), number 5 will be displayed, and `delay(1000)` function of the statement (62) will generate a delay of 1 second.

By using the `digitalWrite` function from statement (63) to statement (69), number 6 will be displayed, and `delay(1000)` function of the statement (70) will generate a delay of 1 second.

By using the `digitalWrite` function from statements (71) to (77), number 7 will be displayed, and `delay(1000)` function of the statement (78) will generate a delay of 1 second.

By using the `digitalWrite` function from statements (79) to (85), number 8 will be displayed, and `delay(1000)` function of the statement (86) will generate a delay of 1 second.

By using the `digitalWrite` function from statements (87) to (93), number 9 will be displayed, and `delay(1000)` function of the statement (94) will generate a delay of 1 second.

Program 4.13

Interface a push-button switch and CC-type seven-segment display with Arduino UNO board, and write a program to display numbers from 0 to 9 in the sequence when the switch is pressed.

int a = 7;	*statement (1)*
int b = 8;	*statement (2)*
int c = 9;	*statement (3)*
int d = 10;	*statement (4)*
int e = 11;	*statement (5)*
int f = 13;	*statement (6)*
int g = 12;	*statement (7)*
void setup()	
{	
pinMode(a,OUTPUT);	*statement (8)*
pinMode(b,OUTPUT);	*statement (9)*
pinMode(c,OUTPUT);	*statement (10)*
pinMode(d,OUTPUT);	*statement (11)*
pinMode(e,OUTPUT);	*statement (12)*
pinMode(f,OUTPUT);	*statement (13)*
pinMode(g,OUTPUT);	*statement (14)*
}	
void loop()	
{	
digitalWrite(a,1);	*statement (15)*
digitalWrite(b,1);	*statement (16)*
digitalWrite(c,1);	*statement (17)*
digitalWrite(d,1);	*statement (18)*
digitalWrite(e,1);	*statement (19)*
digitalWrite(f,1);	*statement (20)*
digitalWrite(g,0); //display 0	*statement (21)*
delay(1000);	*statement (22)*
digitalWrite(a,0);	*statement (23)*
digitalWrite(b,1);	*statement (24)*
digitalWrite(c,1);	*statement (25)*
digitalWrite(d,0);	*statement (26)*
digitalWrite(e,0);	*statement (27)*
digitalWrite(f,0);	*statement (28)*
digitalWrite(g,0); //display 1	*statement (29)*
delay(1000);	*statement (30)*
digitalWrite(a,1);	*statement (31)*
digitalWrite(b,1);	*statement (32)*
digitalWrite(c,0);	*statement (33)*
digitalWrite(d,1);	*statement (34)*
digitalWrite(e,1);	*statement (35)*
digitalWrite(f,0);	*statement (36)*
digitalWrite(g,1); //display 2	*statement (37)*
delay(1000);	*statement (38)*
digitalWrite(a,1);	*statement (39)*
digitalWrite(b,1);	*statement (40)*
digitalWrite(c,1);	*statement (41)*
digitalWrite(d,1);	*statement (42)*
digitalWrite(e,0);	*statement (43)*
digitalWrite(f,0);	*statement (44)*
digitalWrite(g,1); //display 3	*statement (45)*
delay(1000);	*statement (46)*
digitalWrite(a,0);	*statement (47)*

FIGURE 4.26 An Arduino UNO program to display numbers from 0 to 9 in a seven-segment display for the circuit diagram shown in Figure 4.25.

(Continued)

digitalWrite(b,1);	*statement (48)*
digitalWrite(c,1);	*statement (49)*
digitalWrite(d,0);	*statement (50)*
digitalWrite(e,0);	*statement (51)*
digitalWrite(f,1);	*statement (52)*
digitalWrite(g,1); //display 4	*statement (53)*
delay(1000);	*statement (54)*
digitalWrite(a,1);	*statement (55)*
digitalWrite(b,0);	*statement (56)*
digitalWrite(c,1);	*statement (57)*
digitalWrite(d,1);	*statement (58)*
digitalWrite(e,0);	*statement (59)*
digitalWrite(f,1);	*statement (60)*
digitalWrite(g,1); //display 5	*statement (61)*
delay(1000);	*statement (62)*
digitalWrite(a,1);	*statement (63)*
digitalWrite(b,0);	*statement (64)*

digitalWrite(c,1);	*statement (65)*
digitalWrite(d,1);	*statement (66)*
digitalWrite(e,1);	*statement (67)*
digitalWrite(f,1);	*statement (68)*
digitalWrite(g,1); //display 6	*statement (69)*
delay(1000);	*statement (70)*
digitalWrite(a,1);	*statement (71)*
digitalWrite(b,1);	*statement (72)*
digitalWrite(c,1);	*statement (73)*
digitalWrite(d,0);	*statement (74)*
digitalWrite(e,0);	*statement (75)*
digitalWrite(f,0);	*statement (76)*
digitalWrite(g,0); //display 7	*statement (77)*
delay(1000);	*statement (78)*
digitalWrite(a,1);	*statement (79)*
digitalWrite(b,1);	*statement (80)*
digitalWrite(c,1);	*statement (81)*
digitalWrite(d,1);	*statement (82)*
digitalWrite(e,1);	*statement (83)*
digitalWrite(f,1);	*statement (84)*
digitalWrite(g,1); //display 8	*statement (85)*
delay(1000);	*statement (86)*
digitalWrite(a,1);	*statement (87)*
digitalWrite(b,1);	*statement (88)*
digitalWrite(c,1);	*statement (89)*
digitalWrite(d,0);	*statement (90)*
digitalWrite(e,0);	*statement (91)*
digitalWrite(f,1);	*statement (92)*
digitalWrite(g,1); //display 9	*statement (93)*
delay(1000);	*statement (94)*
}	

FIGURE 4.26 (*CONTINUED*) An Arduino UNO program to display numbers from 0 to 9 in a seven-segment display for the circuit diagram shown in Figure 4.25.

Solution

The interfacing of push-button switch and CC-type seven-segment display with Arduino UNO board is shown in Figure 4.27, and the pin-to-pin mapping of CC seven-segment display and Arduino UNO board is shown in Table 4.3. The segments a, b, c, d, e, f, and g of the seven-segment display are connected to pin numbers 7, 8, 9, 10, 11, 13, and 12 of Arduino board through 250 Ω resistor. The com (common) pin of the seven-segment display is connected to GND (ground) pin of Arduino board. The Terminal T2 of the push-button is connected to the GND (ground) pin of Arduino board, and the Terminal T1 is connected to the one terminal of 1 KΩ resistor. The other terminal of 1 KΩ resistor is connected to the 5 V pin of Arduino board. The junction of Terminal T1 of switch and 1 KΩ resistor is extended and connected to the pin number 2 of Arduino board.

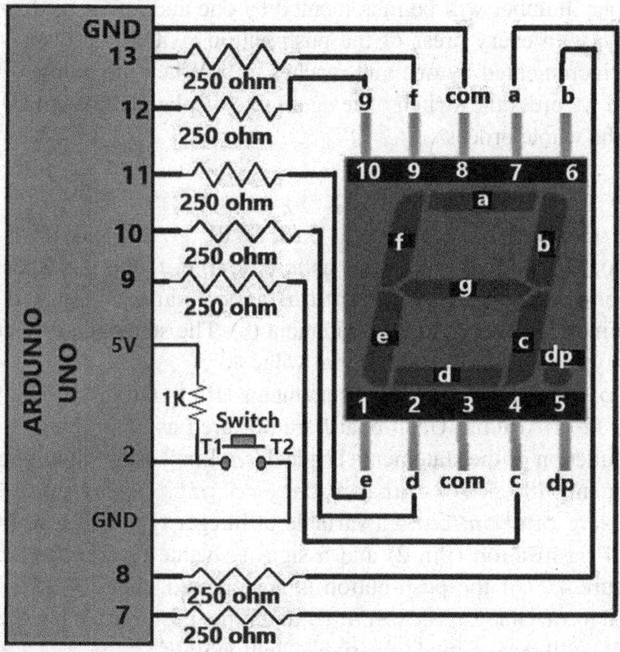

FIGURE 4.27 Interfacing of common cathode seven-segment display and push-button switch with Arduino UNO board.

TABLE 4.3
Pin-to-Pin Mapping of Seven-Segment Display and Push-Button Switch with Arduino UNO Board

CC Seven-Segment Display Pin	Arduino UNO Pin Number/Name
a	7
b	8
c	9
d	10
e	11
f	13
g	12
Common (com)	GND
dp	No connection
Push-button switch	2

An Arduino UNO program for the circuit diagram shown in Figure 4.27 to display numbers from 0 to 9 in a seven-segment display in the sequence when the switch is pressed is shown in Figure 4.28.

The expected operation of the circuit and program shown in Figures 4.27 and 4.28, respectively, is as follows:

Initially, at the reset or power-on of the circuit, the 0 will be displayed on a seven-segment display. If we press the push-button switch once again, then the display number will be incremented by one and 1 will be displayed. In this way, with every press of the push-button switch, the display number will be incremented by one and reaches to 9. When the number displayed is 9, and we press the switch once again, the display will reset to 0 and will repeat the whole process.

Description of the Program:

Using the statements (1) to (7), Pins 7, 8, 9, 10, 11, 13, and 12 of Arduino UNO board are given the names as a, b, c, d, e, f, and g, respectively, of the seven-segment display. The `pushButton` name is assigned to Pin 2 of Arduino UNO board by the statement (8). The statement (9) declares an integer-type variable "p" with initial value "0".

Using the `pinMode` function from statements (10) to (16), Pins 7, 8, 9, 10, 11, 13, and 12 of Arduino UNO board are declared as output pins. The `pinMode` function of the statement (17) declares Pin 2 as an input pin.

The statement (18) `int buttonState=digitalRead(pushButton)` will declare `buttonState` a variable of integer type and read the digital value of pushButton (Pin 2) and assign it's value to `buttonState`. As per Figure 4.27 if the push-button is not pressed, then "1" (5 V) will be assigned to variable `buttonState`, and if the push-button is pressed, then "0" (0 V) will be assigned to variable "buttonState".

The value of `buttonState` is evaluated in the statement (19) `if (buttonState==0)`, i.e., the push-button switch is pressed, then in statement (20) p++ will be executed, and the value of which was initially 0 will be incremented by one and becomes 1.

int a = 7;	*statement (1)*
int b = 8;	*statement (2)*
int c = 9;	*statement (3)*
int d = 10;	*statement (4)*
int e = 11;	*statement (5)*
int f = 13;	*statement (6)*
int g = 12;	*statement (7)*
int pushButton=2;	*statement (8)*
int p=0;	*statement (9)*
void setup()	
{	

FIGURE 4.28 An Arduino UNO program for the circuit diagram shown in Figure 4.27 to display numbers from 0 to 9 in a seven-segment display in the sequence when the switch is pressed.

(Continued)

pinMode(a,OUTPUT);	*statement (10)*
pinMode(b,OUTPUT);	*statement (11)*
pinMode(c,OUTPUT);	*statement (12)*
pinMode(d,OUTPUT);	*statement (13)*
pinMode(e,OUTPUT);	*statement (14)*
pinMode(f,OUTPUT);	*statement (15)*
pinMode(g,OUTPUT);	*statement (16)*
pinMode(pushButton,INPUT);	*statement (17)*
}	
void loop()	
{	
int buttonState=digitalRead(pushButton);	*statement (18)*
if (buttonState==0)//switch is pressed then 0 (LOW) will read	*statement (19)*
{	
p++;	*statement (20)*
}	
if (p==0)	*statement (21)*
{	
digitalWrite(a,1);	*statement (22)*
digitalWrite(b,1);	*statement (23)*
digitalWrite(c,1);	*statement (24)*
digitalWrite(d,1);	*statement (25)*
digitalWrite(e,1);	*statement (26)*
digitalWrite(f,1);	*statement (27)*
digitalWrite(g,0); //display 0	*statement (28)*
delay(1000);	*statement (29)*
}	
if(p==1)	*statement (30)*
{	
digitalWrite(a,0);	*statement (31)*
digitalWrite(b,1);	*statement (32)*
digitalWrite(c,1);	*statement (33)*
digitalWrite(d,0);	*statement (34)*
digitalWrite(e,0);	*statement (35)*
digitalWrite(f,0);	*statement (36)*
digitalWrite(g,0); //display 1	*statement (37)*
delay(1000);	*statement (38)*
}	
if(p==2)	*statement (39)*
{	
digitalWrite(a,1);	*statement (40)*
digitalWrite(b,1);	*statement (41)*
digitalWrite(c,0);	*statement (42)*
digitalWrite(d,1);	*statement (43)*
digitalWrite(e,1);	*statement (44)*
digitalWrite(f,0);	*statement (45)*
digitalWrite(g,1); //display 2	*statement (46)*
delay(1000);	*statement (47)*
}	
if(p==3)	*statement (48)*
{	
digitalWrite(a,1);	*statement (49)*

FIGURE 4.28 (*CONTINUED*) An Arduino UNO program for the circuit diagram shown in Figure 4.27 to display numbers from 0 to 9 in a seven-segment display in the sequence when the switch is pressed.

(Continued)

digitalWrite(b,1);	*statement (50)*
digitalWrite(c,1);	*statement (51)*
digitalWrite(d,1);	*statement (52)*
digitalWrite(e,0);	*statement (53)*
digitalWrite(f,0);	*statement (54)*
digitalWrite(g,1); //display 3	*statement (55)*
delay(1000);	*statement (56)*
}	
if(p==4)	*statement (57)*
{	
digitalWrite(a,0);	*statement (58)*
digitalWrite(b,1);	*statement (59)*
digitalWrite(c,1);	*statement (60)*
digitalWrite(d,0);	*statement (61)*
digitalWrite(e,0);	*statement (62)*
digitalWrite(f,1);	*statement (63)*
digitalWrite(g,1); //display 4	*statement (64)*
delay(1000);	*statement (65)*
}	
if(p==5)	*statement (66)*
{	
digitalWrite(a,1);	*statement (67)*
digitalWrite(b,0);	*statement (68)*
digitalWrite(c,1);	*statement (69)*
digitalWrite(d,1);	*statement (70)*
digitalWrite(e,0);	*statement (71)*
digitalWrite(f,1);	*statement (72)*
digitalWrite(g,1); //display 5	*statement (73)*
delay(1000);	*statement (74)*
}	
if(p==6)	*statement (75)*
{	
digitalWrite(a,0);	*statement (76)*
digitalWrite(b,0);	*statement (77)*
digitalWrite(c,1);	*statement (78)*
digitalWrite(d,1);	*statement (79)*
digitalWrite(e,1);	*statement (80)*
digitalWrite(f,1);	*statement (81)*
digitalWrite(g,1); //display 6	*statement (82)*
delay(1000);	*statement (83)*
}	
if(p==7)	*statement (84)*
{	
digitalWrite(a,1);	*statement (85)*
digitalWrite(b,1);	*statement (86)*
digitalWrite(c,1);	*statement (87)*
digitalWrite(d,0);	*statement (88)*
digitalWrite(e,0);	*statement (89)*
digitalWrite(f,0);	*statement (90)*
digitalWrite(g,0); //display 7	*statement (91)*
delay(1000);	*statement (92)*
}	
if(p==8)	*statement (93)*

FIGURE 4.28 (*CONTINUED*) An Arduino UNO program for the circuit diagram shown in Figure 4.27 to display numbers from 0 to 9 in a seven-segment display in the sequence when the switch is pressed.

(Continued)

{	
digitalWrite(a,1);	statement (94)
digitalWrite(b,1);	statement (95)
digitalWrite(c,1);	statement (96)
digitalWrite(d,1);	statement (97)
digitalWrite(e,1);	statement (98)
digitalWrite(f,1);	statement (99)
digitalWrite(g,1); //display 8	statement (100)
delay(1000);	statement(101)
}	
if(p==9)	statement (102)
{	
digitalWrite(a,1);	statement (103)
digitalWrite(b,1);	statement (104)
digitalWrite(c,1);	statement (105)
digitalWrite(d,0);	statement (106)
digitalWrite(e,0);	statement (107)
digitalWrite(f,1);	statement (108)
digitalWrite(g,1); //display 9	statement (109)
delay(1000);	statement(110)
}	
if(p==10)	statement (111)
{	
p=0;	statement (112)
digitalWrite(a,1);	statement (113)
digitalWrite(b,1);	statement (114)
digitalWrite(c,1);	statement (115)
digitalWrite(d,1);	statement (116)
digitalWrite(e,1);	statement (117)
digitalWrite(f,1);	statement (118)
digitalWrite(g,0); //display 0	statement(119)
delay(1000);	statement(120)
}	
}	

FIGURE 4.28 (*CONTINUED*) An Arduino UNO program for the circuit diagram shown in Figure 4.27 to display numbers from 0 to 9 in a seven-segment display in the sequence when the switch is pressed.

Now statements (21), (30), (39), (48), (57), (66), (75), (84), (93), (102), and (111) will be evaluated. After evaluation, the statement (30) will be satisfied, and statements (31) to (38) will be executed to display 1 and generate a delay of 1 second. In this way, numbers from (0 to 9) are displayed in the sequence when the switch is pressed.

Program 4.14

Re-write the program written in Figure 4.28 by using "for loop" to display numbers from 0 to 9 in a seven-segment display in the sequence when the switch is pressed.

Solution

Refer to Figure 4.27 for interfacing circuit and Table 4.3 for pin-to-pin mapping of CC seven-segment display and push button switch with Arduino UNO board. The program to display numbers from (0 to 9) when the switch is pressed by using `for loop` is shown in Figure 4.29.

int a = 7;	*statement (1)*
int b = 8;	*statement (2)*
int c = 9;	*statement (3)*
int d = 10;	*statement (4)*
int e = 11;	*statement (5)*
int f = 12;	*statement (6)*
int g = 13;	*statement (7)*
int pushButton=2;	*statement (8)*
int p=0;	*statement (9)*
void setup()	
{	
for (int i=7; i<14; i++)	*statement (10)*
{	
pinMode(i,OUTPUT);	*statement (11)*
}	
pinMode(pushButton,INPUT);	*statement (12)*
}	
void loop()	
{	
int buttonState=digitalRead(pushButton);	*statement (13)*
if (buttonState==0)	*statement (14)*
{	
p++;	*statement (15)*
}	
if (p==0)	*statement (16)*
{	
for (int i=7; i<13; i++)	*statement (17)*
{	
digitalWrite(i,1);	*statement (18)*
}	
digitalWrite(g,0); //display 0	*statement (19)*
delay(1000);	*statement (20)*
}	
if(p==1)	*statement (21)*
{	
digitalWrite(a,0);	*statement (22)*
for (int i=8; i<10; i++)	*statement (23)*
{	
digitalWrite(i,1);	*statement (24)*
}	
for (int i=10; i<14; i++)	*statement (25)*
{	
digitalWrite(i,0); //display 1	*statement (26)*
}	
delay(1000);	*statement (27)*
}	
if(p==2)	*statement (28)*
{	
for (int i=7; i<9; i++)	*statement (29)*
{	
digitalWrite(i,1);	*statement (30)*
}	
digitalWrite(c,0);	*statement (31)*

FIGURE 4.29 An Arduino UNO program by using loop for the circuit diagram shown in Figure 4.27 to display numbers from 0 to 9 in a seven-segment display in the sequence when the switch is pressed.

(Continued)

for (int i=10; i<12; i++)	*statement (32)*
{	
digitalWrite(i,1);	*statement (33)*
}	
digitalWrite(f,0);	*statement (34)*
digitalWrite(g,1); //display 2	*statement (35)*
delay(1000);	*statement (36)*
}	
if(p==3)	*statement (37)*
{	
for (int i=7; i<11; i++)	*statement (38)*
{	
digitalWrite(i,1);	*statement (39)*
}	
for (int i=11; i<13; i++)	*statement (40)*
{	
digitalWrite(i,0);	*statement (41)*
}	
digitalWrite(g,1); //display 3	*statement (42)*
delay(1000);	*statement (43)*
}	
if(p==4)	*statement (44)*
{	
digitalWrite(a,0);	*statement (45)*
for (int i=8; i<10; i++)	*statement (46)*
{	
digitalWrite(i,1);	*statement (47)*
}	
for (int i=10; i<12; i++)	*statement (48)*
{	
digitalWrite(i,0);	*statement (49)*
}	
for (int i=12; i<14; i++)	*statement (50)*
{	
digitalWrite(i,1); //display 4	*statement (51)*
}	
delay(1000);	*statement (52)*
}	
if(p==5)	*statement (53)*
{	
digitalWrite(a,1);	*statement (54)*
digitalWrite(b,0);	*statement (55)*
for (int i=9; i<11; i++)	*statement (56)*
{	
digitalWrite(i,1);	*statement (57)*
}	
digitalWrite(e,0);	*statement (58)*
for (int i=12; i<14; i++)	*statement (59)*
{	
digitalWrite(i,1); //display 5	*statement (60)*
}	
delay(1000);	*statement (61)*
}	

FIGURE 4.29 (*CONTINUED*) An Arduino UNO program by using loop for the circuit diagram shown in Figure 4.27 to display numbers from 0 to 9 in a seven-segment display in the sequence when the switch is pressed.

(Continued)

if(p==6)	*statement (62)*
{	
for (int i=7; i<9; i++)	*statement (63)*
{	
digitalWrite(i,0);	*statement (64)*
}	
for (int i=9; i<14; i++)	*statement (65)*
{	
digitalWrite(i,1); //display 6	*statement (66)*
}	
delay(1000);	*statement (67)*
}	
if(p==7)	*statement (68)*
{	
for (int i=7; i<10; i++)	*statement (69)*
{	
digitalWrite(i,1);	*statement (70)*
}	
for (int i=10; i<14; i++)	*statement (71)*
{	
digitalWrite(i,0); //display 7	*statement (72)*
}	
delay(1000);	*statement (73)*
}	
if(p==8)	*statement (74)*
{	
{	
for (int i=7; i<14; i++)	*statement (75)*
{	
digitalWrite(i,1); //display 8	*statement (76)*
}	
delay(1000);	*statement (77)*
}	
if(p==9)	*statement (78)*
{	
for (int i=7; i<10; i++)	*statement (79)*
{	
digitalWrite(i,1);	*statement (80)*
}	
for (int i=10; i<12; i++)	*statement (81)*
{	
digitalWrite(i,0);	*statement (82)*
}	
for (int i=12; i<14; i++)	*statement (83)*
{	
digitalWrite(i,1); //display 9	*statement (84)*
}	
delay(1000);	*statement (85)*
}	
if(p==10)	*statement (86)*
{	
p=0;	*statement (87)*
for (int i=7; i<13; i++)	*statement (88)*

FIGURE 4.29 (*CONTINUED*) An Arduino UNO program by using loop for the circuit diagram shown in Figure 4.27 to display numbers from 0 to 9 in a seven-segment display in the sequence when the switch is pressed.

(Continued)

{	
digitalWrite(i,1);	statement (89)
}	
digitalWrite(g,0); //display 0	statement (90)
delay(1000);	statement (91)
}	
}	

FIGURE 4.29 (*CONTINUED*) An Arduino UNO program by using loop for the circuit diagram shown in Figure 4.27 to display numbers from 0 to 9 in a seven-segment display in the sequence when the switch is pressed.

Description of the Program:

By using the statement (10) `for (int i=7; i<14; i++)`, an integer i is created with an initial value 7. The condition to be tested for running the loop is i<14. Whenever i is less than 14, the statement (11) `pinMode(i, OUTPUT)` will be executed, and Pins 7–13 will be initialized as an output pin. After initializing Pin 13 as an output pin, the `loop` will end, and statement (12) `pinMode(pushButton, INPUT)` will be executed and declares Pin 2 as an input pin. The statements inside the `void loop()` are self-explanatory.

4.5 MISCELLANEOUS PROGRAMS RELATED TO LED

In this section, we shall discuss some programs based on arrays and loop to generate different types of patterns using LEDs. The working principle of LED is explained in Section 3.1 of Chapter 3.

Program 4.15

Interface seven LEDs with Arduino UNO board, and write a program using an array to on them in a sequence and then off them in the same sequence.

Solution

The interfacing of seven LEDs with Arduino UNO board is shown in Figure 4.30. The anode of seven LEDs is connected to the pin numbers 7–13 of Arduino UNO board through a 250 Ω resistor, and the cathode is connected to the GND (ground) pin of Arduino board. An Arduino UNO program for the circuit diagram shown in Figure 4.30 to on seven LEDs in a sequence and then off them in the same sequence using array is shown in Figure 4.31.

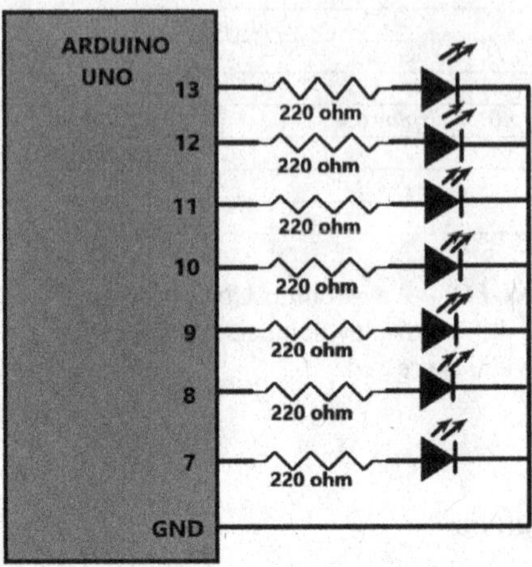

FIGURE 4.30 Interfacing of seven LEDs with Arduino UNO board.

int led_array[7]={7,8,9,10,11,12,13};	*statement (1)*
void setup()	
{	
for (int i=0; i<=6; i++)	*statement (2)*
{	
pinMode(led_array[i],OUTPUT);	*statement (3)*
}	
}	
void loop()	
{	
for (int i=0; i<=6; i++)	*statement (4)*
{	
digitalWrite(led_array[i],HIGH);	*statement (5)*
delay(200);	*statement (6)*
}	
for (int i=0; i<=6; i++)	*statement (7)*
{	
digitalWrite(led_array[i],LOW);	*statement (8)*
delay(200);	*statement (9)*
}	
}	

FIGURE 4.31 An Arduino UNO program for the circuit diagram shown in Figure 4.30 to on seven LEDs in a sequence and then off them in the same sequence using array.

The expected operation of the circuit and program shown in Figures 4.30 and 4.31, respectively, is as follows:

> Initially, all the seven LEDs from D1 to D7 are off at the reset or power-on of the circuit. At first, LED D1 will on and subsequently after every 200 ms LEDs from D2 to D7 will on in the sequence. Once all LEDs are on, LED D1 will off and subsequently after every 200 ms LEDs from D2 to D7 will off in the sequence. The above-described sequence will repeat as long as power is on.

Description of the Program:

The statement (1) *int led _ array[7]={7,8,9,10,11,12,13}* is used to create an array of seven elements. The seven elements of the array can be accessed as led_array[0], led_array[1], led_array[2], led_array[3], led_array[4], led_array[5], and led_array[6].

The arrays elements from led_array[0] to led_array[6] are assigned the Pins (7–13), respectively.

Due to the statement (2) *for (int i=0; i<=6; i++)* and statement (3) *pinMode(led _ array[i], OUTPUT)* led_array[0] to led_array[6], i.e., pin number 7–13 are declared as output pins.

Due to the statement (4) *for (int i=0; i<=6; i++)*, statement (5) *digitalWrite(led _ array[i], HIGH)* and statement (6) *delay(200)* All the LEDs connected from Pin 7 to 13 will on one after other in sequence after every 200 ms.

Due to the statement (7) *for (int i=0; i<=6; i++)*, statement (8) *digitalWrite(led _ array[i], LOW)*, and statement (9) *delay(200)*, all the LEDs connected from Pin 7 to 13 will off one after other in sequence after every 200 ms.

Program 4.16

Interface seven LEDs with Arduino UNO board as shown in Figure 4.30, and write a program using an array to on them in a sequence and then off them in the reverse sequence.

Solution

The interfacing of seven LEDs with Arduino UNO board is shown in Figure 4.30. The cathode terminal of all LEDs is connected to GND pin of Arduino UNO, and the anode terminal of each LED is connected to Pin from (7 to 13) through a 250 Ω register. An Arduino UNO program for the circuit diagram shown in Figure 4.30 to on seven LEDs in a sequence and then off them in the reverse sequence using array is shown in Figure 4.32.

The Expected Operation of the circuit and program shown in Figures 4.30 and 4.32, respectively, is as follows:

Initially, all the seven LEDs connected from Pin 7 to 13 are off. First, LED connected at Pin 7 will on and subsequently after every 200 ms LEDs connected from Pin 8 to 13 will on in the sequence. Once all LEDs are on, LED at Pin 13 will off and subsequently after every 200 ms, LEDs connected from Pin 2 to 7 will off in the sequence. The above-described sequence will repeat as long as power is on.

Description of the Program:

The statement (1) int *led _ array[7]={7,8,9,10,11,12,13}* is used to create an array of seven elements. The seven elements of the array can be accessed as led_array[0], led_array[1], led_array[2], led_array[3], led_array[4], led_array[5], and led_array[6].

int led_array[7]={7,8,9,10,11,12,13};	*statement (1)*
void setup()	
{	
for (int i=0; i<=6; i++)	*statement (2)*
{	
pinMode(led_array[i],OUTPUT);	*statement (3)*
}	
}	
void loop()	
{	
for (int i=0; i<=6; i++)	*statement (4)*
{	
digitalWrite(led_array[i],HIGH);	*statement (5)*
delay(200);	*statement (6)*
}	
for (int i=7; i>=0; i--)	*statement (7)*
{	
digitalWrite(led_array[i],LOW);	*statement (8)*
delay(200);	*statement (9)*
}	
}	

FIGURE 4.32 An Arduino UNO program for the circuit diagram shown in Figure 4.30 to on seven LEDs in a sequence and then off them in the reverse sequence using array.

The arrays elements from led_array[0] to led_array[6] are assigned the Pin from (7 to 13), respectively.

Due to the statement (2) `for (int i=0; i<=6; i++)` and statement (3) `pinMode(led _ array[i], OUTPUT)` led_array[0] to led_array[6], i.e., Pins (7–13) are declared as output pins.

Due to the statement (4) for `(int i=0; i<=6; i++)`, statement (5) `digitalWrite(led _ array[i], HIGH)` and statement (6) `delay(200)` All the LEDs connected from Pin 7 to 13 will on one after other in sequence after every 200 ms.

Due to the statement (7) `for (int i=7; i>=0; i--)`, statement (8) `digitalWrite(led _ array[i], LOW)` and statement (9) `delay(200)`, all the LEDs connected from Pin 13 to 7 will off one after other in sequence after every 200 ms.

Program 4.17

Interface a LED with Arduino UNO board, and write a program to toggle it for five times.

Solution

The interfacing LED with Arduino UNO board is shown in Figure 4.33. The anode of LED is connected to the pin number 2 of Arduino UNO board through a 250 Ω resistor, and the cathode is connected to the GND (ground) pin of Arduino board. An Arduino UNO program for the circuit diagram shown in Figure 4.33 to toggle a LED for five times is shown in Figure 4.34.

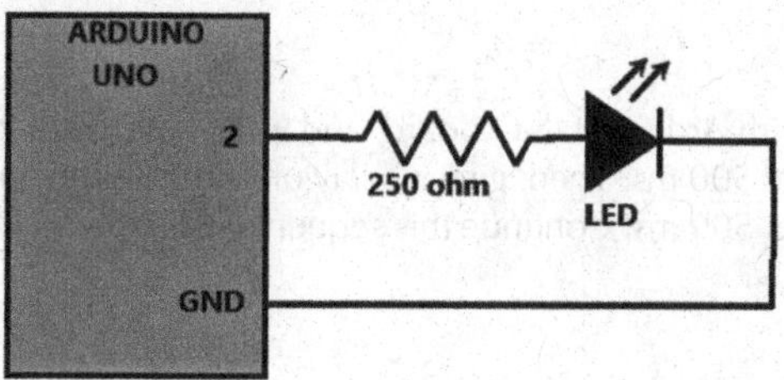

FIGURE 4.33 Interfacing of a LED with Arduino UNO board for Program 4.17.

int led=2;	*statement (1)*
void setup()	
{	
pinMode(led,OUTPUT);	*statement (2)*
for(int i=0;i<=4;i++)	*statement (3)*
{	
digitalWrite(led,HIGH);	*statement (4)*
delay(500);	*statement (5)*
digitalWrite(led,LOW);	*statement (6)*
delay(500);	*statement (7)*
}	
}	
void loop()	
{	
}	

FIGURE 4.34 An Arduino UNO program for the circuit diagram shown in Figure 4.33 to toggle a LED for five times.

The Expected Operation of the circuit and program shown in Figures 4.33 and 4.34, respectively, is as follows:

When the circuit is on the LED will on, and after 500 ms, it will be off, and this sequence will be repeated for five times.

Description of the Program:

The statement (1) `int led=2` is used to assign name LED to Pin 2.

The statement (2) `pinMode(led, OUTPUT)` is used to declare Pin 2 as an output pin.

The statements (4), (5), (6), and (7) will run five times and thus make the LED on and off with a gap of 500 ms five times.

The statements (4), (5), (6), and (7) will run five times because statement (3) `for (int i=0; i<=4; i++)` will allow the `for loop` to run five times only.

Since the statements (2), (3), (4), (5), (6), and (7) are placed inside void `setup()`, we know that the statements inside `setup()` will run only once therefore, the LED will on and off with a gap of 500 ms five times only.

Program 4.18

Interface a LED with Arduino UNO board, and write a program to toggle it for five times with a gap of 500 ms. Then, give a delay of 3 seconds and again toggle it five times with a gap of 500 ms. Continue this sequence till Arduino UNO is power-on.

Solution

The interfacing LED with Arduino UNO board is shown in Figure 4.33. The anode of LED is connected to the pin number 2 of Arduino UNO board through a 250 Ω resistor, and the cathode is connected to the GND (ground) pin of Arduino board. An Arduino UNO program for the circuit diagram shown in Figure 4.33 to toggle a LED for five times with a gap of 500 ms and continue this sequence till Arduino UNO is power on with a gap of 3 seconds is shown in Figure 4.35.

The expected operation of the circuit and program shown in Figures 4.33 and 4.35, respectively, is as follows:

> When the circuit is on the LED will on, and after 500 ms, it will be off, and this sequence will be repeated five times. Then after a delay of 3 seconds, the same on/off sequence will repeat. And this sequence will continue as long as power is on.

Description of the Program:

The statement (1) `int led=2` is used to assign name LED to Pin 2.
The statement (2) `pinMode(led, OUTPUT)` is used to declare Pin 2 as an output pin.

int led=2;	*statement (1)*
void setup()	
{	
pinMode(led,OUTPUT);	*statement (2)*
for(int a=0;a<=2;a++)	*statement (3)*
{	
for(int i=0;i<=4;i++)	*statement (4)*
{	
digitalWrite(led,HIGH);	*statement (5)*
delay(500);	*statement (6)*
digitalWrite(led,LOW);	*statement (7)*
delay(500);	*statement (8)*
}	
Delay(3000)	*statement (9)*
}	
}	
void loop()	
{	
}	

FIGURE 4.35 An Arduino UNO program for the circuit diagram shown in Figure 4.33 to toggle a LED for five times with a gap of 500 ms, and continue this sequence till Arduino UNO is power on with a gap of 3 seconds.

The statements (5), (6), (7), and (8) will run five times and thus make the LED on and off with a gap of 500 ms five times because of the statement (4) `for (int i=0; i<=4; i++)`.

The statements from (4) to (9) will run three times with a gap of 3 seconds due to statement (3) because statement (3) `for (int i=0; i<=4; i++)` will allow the "for loop" to run three times only.

Since all the statements are placed inside the `void setup()`, we know that the statements inside `setup()` will run only once. Therefore, the LED will toggle for five times with a gap of 500 ms and continue this sequence till Arduino UNO is power on with a gap of 3 seconds.

4.6 LCD INTERFACING AND PROGRAMMING

This section shall discuss the interfacing of LCD with Arduino UNO board and various programs to display characters in LCD. In this section, readers will also understand how to scroll the displayed message on LCD. A 16×2 LCD is used for the demonstration of programs in this section. The working principle of LCD is explained in Section 3.4 of Chapter 3.

Program 4.19

Interface a LCD with Arduino UNO board, and write a program to display "YOGESH MISRA" from the 0th row and 0th column.

Solution

The interfacing of LCD with Arduino UNO board and the pin-to-pin mapping of LCD and Arduino UNO board are shown in Figure 4.36 and Table 4.4, respectively. The VDD and +5 V pin of LCD is connected to the 5 V pin of Arduino board. The VSS, GND, and RD/WR′ (Read/Write′) pin of LCD is connected to the GND (ground) pin of Arduino board. The RS (Register Select) pin of LCD is connected to the pin number 2 of Arduino board. The EN (Enable) pin of LCD is connected to the pin number 3 of Arduino board. The Terminals T1 and T2 of 5 KΩ potentiometer are connected to the 5 V and GND (ground) pin of Arduino board respectively. The VEE pin of LCD is connected to the middle (wiper) terminal of 5 KΩ potentiometer. The D4–D7 pins of LCD are connected to the pin numbers 4–7 of Arduino board.

A program to display "YOGESH MISRA" from 0th row and 0th column is shown in Figure 4.37. The display of "YOGESH MISRA" from the 0th row and 0th column in LCD after the execution of Arduino program is shown in Figure 4.38.

Description of the Program:

The statement (1) `#include <LiquidCrystal.h>` is used to include LCD library. The statement (2) `int RS=2, EN=3, D4=4, D5=5, D6=6, D7=7` initializes LCD pins with Arduino UNO pins as shown in Table 4.5.

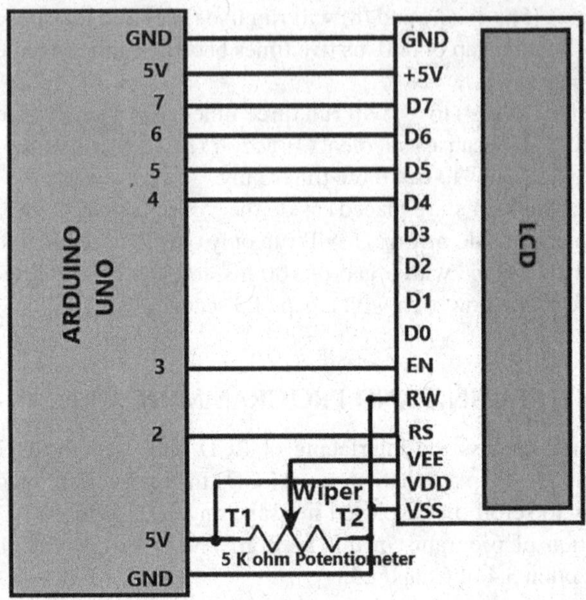

FIGURE 4.36 Interfacing of LCD display with Arduino UNO board.

TABLE 4.4

Pin-to-Pin Mapping of LCD Display with Arduino UNO Board of Program 4.19

LCD Display Pin Number	LCD Display Pin Name	Arduino UNO Pin Number/Name
1	VSS (Ground)	Gnd
2	VDD (+5 V)	+5 V
3	VEE (wiper pin of potentiometer)	-
4	Register Select (RS)	2
5	RD/WR′	Gnd
6	Enable (EN)	3
7–10	D0–D3	-
11–14	D4–D7	4–7
15	+5 V (backlight)	+5 V
16	Gnd (backlight)	Gnd

#include <LiquidCrystal.h>	*statement (1)*
int RS=2, EN=3, D4=4, D5=5, D6=6, D7=7;	*statement (2)*
LiquidCrystal lcd(RS, EN, D4, D5, D6, D7);	*statement (3)*
void setup()	
{	
lcd.begin(16, 2);	*statement (4)*
}	
void loop()	
{	
lcd.setCursor(0,0);	*statement (5)*
lcd.print("YOGESH MISRA");	*statement (6)*
}	

FIGURE 4.37 An Arduino UNO program to display "YOGESH MISRA" on LCD from the 0th row and 0th column.

TABLE 4.5
Pin-to-Pin Mapping of LCD Display and "lcd" Object Created in Statement (3) of Program Shown in Figure 4.37

LCD Display Pin Name	Arduino UNO Pin Number/Name
RS (Register Select)	2
EN (Enable)	3
D4	4
D5	5
D6	6
D7	7

FIGURE 4.38 The display of "YOGESH MISRA" from the 0th row and 0th column in LCD after the execution of Arduino program shown in Figure 4.37.

The statement (3) `LiquidCrystal lcd(RS, EN, D4, D5, D6, D7)` creates an object "lcd" with pin names (RS, EN, D4, D5, D6, D7), and statement (2) has already assigned the pin names of "lcd" object as shown in Table 4.5.

The statement (4) `lcd.begin(16,2)` is used to initialize the "lcd" object created in statement (3) as a 16 column, 2 row LCD display.

The statement (5) is used to initialize from where the display of characters will start. The `lcd.setCursor(0,0)` statement initializes the display from 0th column and 0th row.

The statement (6) `lcd.print("YOGESH MISRA")` will display "YOGESH MISRA" on LCD display.

Program 4.20

Interface a LCD display with Arduino UNO board, and write a program to display "YOGESH MISRA" from the 0th row and 2nd column.

Solution

The interfacing and the pin-to-pin mapping of LCD and Arduino UNO board are shown in Figure 4.36 and Table 4.4, respectively. A program to display "YOGESH MISRA" from the 0th row and 2nd column is shown in Figure 4.39.

The display of "YOGESH MISRA" from the 0th row and 2nd column in LCD after the execution of Arduino program is shown in Figure 4.40.

Description of the Program:

The program shown in Figure 4.39 is different from the program shown in Figure 4.37 only in the statement (5).

The statement (5) of program shown in Figure 4.37 was `lcd.setCursor(0,0)` due to which the "YOGESH MISRA" displayed from the 0th column of 0th row. In contrast, the statement (5) of program shown in Figure 4.39 was `lcd.setCursor(2,0)` due to which the "YOGESH MISRA" displayed from 2nd column of 0th row.

#include <LiquidCrystal.h>	*statement (1)*
int RS=2, EN=3, D4=4, D5=5, D6=6, D7=7;	*statement (2)*
LiquidCrystal lcd(RS, EN, D4, D5, D6, D7);	*statement (3)*
void setup()	
{	
lcd.begin(16, 2);	*statement (4)*
}	
void loop()	
{	
lcd.setCursor(2,0);	*statement (5)*
lcd.print("YOGESH MISRA");	*statement (6)*
}	

FIGURE 4.39 An Arduino UNO program to display "YOGESH MISRA" on LCD from the 0th row and 2nd column.

FIGURE 4.40 Display of "YOGESH MISRA" from the 0th row and 2nd column after the execution of program shown in Figure 4.39.

Program 4.21

Interface an LCD with Arduino UNO board, and write a program to display "YOGESH MISRA" from the 0th row and 2nd column and "WELCOMES YOU" from the 1st row and 2nd column.

Solution

The interfacing and the pin-to-pin mapping of LCD and Arduino UNO board are shown in Figure 4.36 and Table 4.4, respectively. A program to display "YOGESH MISRA" from the 0th row and 2nd column and "WELCOMES YOU" from the 1st row and 2nd column is shown in Figure 4.41. After the execution of the program, the displayed result is shown in Figure 4.42.

Description of the Program:

The program was written from statement (1) to statement (6), which will display "YOGESH MISRA" from 2nd column of 0th row.

The statement (7) `lcd.setCursor(2,1)` initializes the display from 2nd column and 1st row, and statement (8) `lcd.print("WELCOMES YOU")` will display "WELCOMES YOU" from 2nd column and 1st row.

After the execution of the program shown in Figure 4.41, the characters will be displayed, as shown in Figure 4.42.

#include <LiquidCrystal.h>	statement (1)
int RS=2, EN=3, D4=4, D5=5, D6=6, D7=7;	statement (2)
LiquidCrystal lcd(RS, EN, D4, D5, D6, D7);	statement (3)
void setup()	
{	
lcd.begin(16,2);	statement (4)
}	
void loop()	
{	
lcd.setCursor(2,0);	statement (5)
lcd.print("YOGESH MISRA");	statement (6)
lcd.setCursor(2,1);	statement (7)
lcd.print("WELCOMES YOU");	statement (8)
}	

FIGURE 4.41 An Arduino UNO program to display "YOGESH MISRA" on LCD from the 0th row and 2nd column and "WELCOMES YOU" from the 1st row and 2nd column.

FIGURE 4.42 Display of "YOGESH MISRA" from the 0th row and 2nd column and "WELCOMES YOU" from the 1st row and 2nd column after the execution of program shown in Figure 4.41.

Program 4.22

Interface an LCD with Arduino UNO board, and write a program to display "YOGESH MISRA" from the 0th row and 0th column and scroll the message right.

Solution

The interfacing and the pin-to-pin mapping of LCD and Arduino UNO board are shown in Figure 4.36 and Table 4.4, respectively. A program to display "YOGESH MISRA" from the 0th row and 0th column and scrolling of the message to the right is shown in Figure 4.43.

Description of the Program:

The display "YOGESH MISRA" from the 0th row and 0th column is implemented by using statement (1) to statement (6), and the description of the program is the same as the description of the program shown in Figure 4.37.

The statement (7) `lcd.scrollDisplayRight()` scrolls the message towards the right side of the LCD.

Program 4.23

Interface an LCD with Arduino UNO board, and write a program to:
(A) Display "YOGESH MISRA" from the 0th row and 0th column for 1 second, clear the display for 1 second, and repeat the sequence.
(B) Display "YOGESH MISRA" from the 0th row and 0th column and scroll it to the right.

Solution

(A) The interfacing and the pin-to-pin mapping of LCD and Arduino UNO board are shown in Figure 4.36 and Table 4.4, respectively. A program to display "YOGESHMISRA" from the 0th row and 0th column for 1 second, clear the display for 1 second, and repeat the sequence is shown in Figure 4.43.

#include <LiquidCrystal.h>	*statement (1)*
int RS=2, EN=3, D4=4, D5=5, D6=6, D7=7;	*statement (2)*
LiquidCrystal lcd(RS, EN, D4, D5, D6, D7);	*statement (3)*
void setup()	
{	
lcd.begin(16,2);	*statement (4)*
}	
void loop()	
{	
lcd.setCursor(0,0);	*statement (5)*
lcd.print("YOGESH MISRA");	*statement (6)*
delay(1000);	*statement (7)*
lcd.clear();	*statement (8)*
delay(1000);	*statement (9)*
}	

FIGURE 4.43 An Arduino UNO program to display "YOGESH MISRA" on LCD from the 0th row and 0th column for 1 second, clear the display for 1 second and repeat the sequence.

Description of the Program:

The display "YOGESH MISRA" from the 0th row and 0th column is implemented by using statement (1) to statement (6), and the description of the program is the same as the description of the program as shown in Figure 4.37.

The statement (7) `delay(1000)` will generate a delay of 1 second, and "YOGESH MISRA" will be displayed for 1 second.

The statement (8) `lcd.clear()` will clear the LCD, and statement (9) "delay(1000)" will again generate a delay of 1 second.

Now `void loop()` function will repeat, and the sequence of the display will be repeated.

(B) A program to display "YOGESH MISRA" from the 0th row and 0th column and scroll it to the right is shown in Figure 4.44. The program is self explanatory. The statement 7 lcd.scrollDisplayRight() will scroll the message in right direction.

#include <LiquidCrystal.h>	*statement (1)*
int RS=2, EN=3, D4=4, D5=5, D6=6, D7=7;	*statement (2)*
LiquidCrystal lcd(RS, EN, D4, D5, D6, D7);	*statement (3)*
void setup()	
{	
lcd.begin(16,2);	*statement (4)*
}	
void loop()	
{	
lcd.setCursor(0,0);	*statement (5)*
lcd.print("YOGESH MISRA");	*statement (6)*
lcd.scrollDisplayRight();	*statement (7)*
}	

FIGURE 4.44 An Arduino UNO program to display "YOGESH MISRA" on LCD from the 0th row and 0th column and scrolling of the message to the right.

Program 4.24

Interface an LCD with Arduino UNO board, and write a program to perform the following steps:

Step 1: Display "YOGESH MISRA" from the 0th row and 0th column for 1 second.
Step 2: Clear the display.
Step 3: Display "WELCOMES YOU" from the 1st row and 0th column for 1 second.
Step 4: Clear the display.
Step 5: Repeat the steps.

Solution

The interfacing and the pin-to-pin mapping of LCD and Arduino UNO board are shown in Figure 4.36 and Table 4.4, respectively. A program to perform the steps as mentioned in the question is shown in Figure 4.45.

#include <LiquidCrystal.h>	*statement (1)*
int RS=2, EN=3, D4=4, D5=5, D6=6, D7=7;	*statement (2)*
LiquidCrystal lcd(RS, EN, D4, D5, D6, D7);	*statement (3)*
void setup()	
{	
lcd.begin(16,2);	*statement (4)*
}	
void loop()	
{	
lcd.setCursor(0,0);	*statement (5)*
lcd.print("Yogesh Misra");	*statement (6)*
delay(1000);	*statement (7)*
lcd.clear();	*statement (8)*
lcd.setCursor(0,1);	*statement (9)*
lcd.print("WELCOMES YOU");	*statement (10)*
delay(1000);	*statement (11)*
lcd.clear();	*statement (12)*
}	

FIGURE 4.45 An Arduino program to perform the steps as given in Program 4.24.

Description of the Program:

The statements from (1) to (4) are used for the initialization.

The statements from (5) to (8) will perform the first two steps, i.e., Step 1: Display "YOGESH MISRA" from the 0th row and 0th column for 1 second and Step 2: Clear the display.

The statements from (9) to (12) will perform the last two steps, i.e., Step 3: Display "WELCOMES YOU" from the 1st row and 0th column for 1 second and Step 4: Clear the display.

Since statements from (5) to (12) are placed inside the `void loop ()`, Step 1 to Step 4 will repeat.

Program 4.25

Interface an LCD with Arduino UNO board, and write a program to perform the following steps:

Step 1: Display "YOGESH MISRA" from the 0th row and 0th column for 1 second.

Step 2: Scroll the message right side at a speed of 800 ms per character until "A" of "YOGESH MISRA" reaches the 15th column.

Step 3: Repeat the steps.

Solution

The interfacing and the pin-to-pin mapping of LCD and Arduino UNO board are shown in Figure 4.36 and Table 4.4, respectively. A program to perform the steps as mentioned in the question is shown in Figure 4.46.

#include <LiquidCrystal.h>	statement (1)
int RS=2, EN=3, D4=4, D5=5, D6=6, D7=7;	statement (2)
LiquidCrystal lcd(RS, EN, D4, D5, D6, D7);	statement (3)
void setup()	
{	
lcd.begin(16,2);	statement (4)
}	
void loop()	
{	
lcd.setCursor(0,0);	statement (5)
lcd.print("Yogesh Misra");	statement (6)
delay(1000);	statement (7)
for (int position = 0; position <4; position++)	statement (8)
{	
lcd.scrollDisplayRight();	statement (9)
delay(800);	statement (10)
}	
lcd.clear();	statement (11)
}	

FIGURE 4.46 An Arduino program to perform the steps as given in Program 4.25.

Description of the Program:

The statements from (1) to (4) are used for the initialization.

The statements from (5) to (7) will perform Step 1: Display "YOGESH MISRA" from the 0th row and 0th column for 1 second.

The statements from (8) to (10) will perform Step 2: Scroll the message right side at a speed of 800 ms per character till "A" of "YOGESH MISRA" reaches to 15th column.

The statements from (11) will clear the display.

Since statements from (5) to (11) are placed inside the `void loop` (), Step 1 to Step 3 will repeat.

Program 4.26

Interface an LCD with Arduino UNO board, and write a program to perform the following steps:

Step 1: Display "YOGESH MISRA" from the 0th row and 0th column for 1 second.

Step 2: Scroll the message right side at a speed of 800 ms per character until "A" of "YOGESH MISRA" reaches the 15th column.

Step 3: Scroll the message left side at a speed of 800 ms per character until "Y" of "YOGESH MISRA" reaches the 0th column.

Step 4: Repeat the steps.

Solution

The interfacing and the pin-to-pin mapping of LCD and Arduino UNO board are shown in Figure 4.36 and Table 4.4, respectively. A program to perform the steps as mentioned in the question is shown in Figure 4.47.

#include <LiquidCrystal.h>	*statement (1)*
int RS=2, EN=3, D4=4, D5=5, D6=6, D7=7;	*statement (2)*
LiquidCrystal lcd(RS, EN, D4, D5, D6, D7);	*statement (3)*
void setup()	
{	
lcd.begin(16,2);	*statement (4)*
}	
void loop()	
{	
lcd.setCursor(0,0);	*statement (5)*
lcd.print("Yogesh Misra");	*statement (6)*
delay(1000);	*statement (7)*
for (int position = 0; position <4; position++)	*statement (8)*
{	
lcd.scrollDisplayRight();	*statement (9)*
delay(800);	*statement (10)*
}	
for (int position = 0; position <4; position++)	*statement (11)*
{	
lcd.scrollDisplayLeft();	*statement (12)*
delay(800);	*statement (13)*
}	
lcd.clear();	*statement (14)*
}	

FIGURE 4.47 An Arduino UNO program to perform the steps as given in Program 4.26.

Description of the Program:

The statements from (1) to (4) are used for the initialization.
The statements from (5) to (7) will perform Step 1: Display "YOGESH MISRA" from the 0th row and 0th column for 1 second.
The statements from (8) to (10) will perform Step 2: Scroll the message right side at a speed of 800 ms per character till "A" of "YOGESH MISRA" reaches to 15th column.
The statements from (11) to (13) will perform Step 3: Scroll the message left side at a speed of 800 ms per character till "Y" of "YOGESH MISRA" reaches to 0th column.
The statements from (14) will clear the display.
Since statements from (5) to (14) are placed inside the `void loop` (),Step 1 to Step 4 will repeat.

4.7 POTENTIOMETER INTERFACING AND PROGRAMMING

This section shall discuss the interfacing of the potentiometer with the Arduino UNO board and various related programs. After going through this section, reader will understand the concept of inputting the analog signal to Arduino UNO board through A0 to A5 analog input pins. This section will be useful for the readers when they learn the interfacing of sensors with the Arduino board because most of the sensors generate analog signal in response to the physical parameters. The working principle of potentiometer and analog-to-digital conversion is explained in Sections 3.5 and 3.6 of Chapter 3.

Program 4.27

Interface a potentiometer with Arduino UNO board, and write a program to display the various steps of internal analog-to-digital converter on the serial monitor when the potentiometer knob is rotated.

Solution

The interfacing of potentiometer with Arduino UNO board is shown in Figure 4.48. The Terminals T1 and T2 of 10 KΩ potentiometer are connected to the 5 V and GND (ground) pin of Arduino board. The wiper terminal of 10 KΩ potentiometer is connected to A0 pin of Arduino board. A program to display the various steps of internal analog-to-digital converter on the serial monitor when the potentiometer knob is rotated is shown in Figure 4.49. The screenshot of serial monitor of program shown in Figure 4.49 and interfacing circuit shown in Figure 4.48 to display the various steps on the serial monitor when the potentiometer knob is rotated is shown in Figure 4.50.

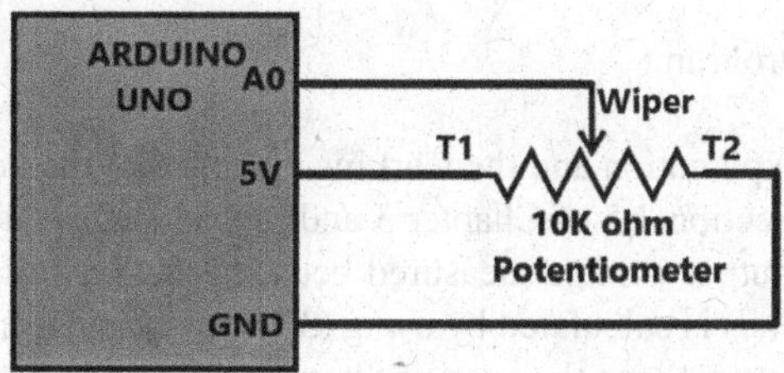

FIGURE 4.48 Interfacing of the potentiometer with Arduino UNO board.

int analogInput=A0;	*statement (1)*
void setup()	
{	
pinMode(analogInput,INPUT);	*statement (2)*
Serial.begin(9600);	*statement (3)*
}	
void loop()	
{	
int step=analogRead(analogInput);	*statement (4)*
Serial.print("Step Number= ");	*statement (5)*
Serial.println(step);	*statement (6)*
delay(1000);	*statement (7)*
}	

FIGURE 4.49 An Arduino UNO program to display the various steps on the serial monitor when the potentiometer knob is rotated for the circuit diagram shown in Figure 4.48.

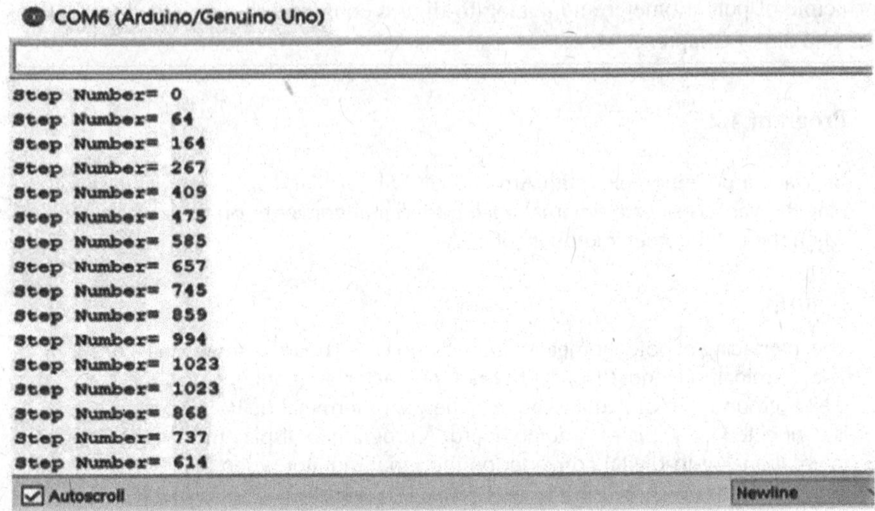

FIGURE 4.50 The screenshot of serial monitor of program shown in Figure 4.49 and interfacing circuit shown in Figure 4.48 to display the various steps on the serial monitor when the potentiometer knob is rotated.

Description of the Program:

The theoretical explanation and the working principle of the potentiometer are explained in Section 3.5 of Chapter 3 and reproduce here for the benefit of readers. The output voltage measured between the Terminal 2 (wiper) and Terminal 3 (Gnd) is calculated by using (3.1) and varies from 0–5 V; thus, it is analog in nature. Since the output voltage of the potentiometer is analog, it can be connected to the analog pins of Arduino. The Arduino board contains six analog pins named A0, A1, A2, A3, A4, and A5. Internally, these analog pins are connected to a six-channel 10-bit analog-to-digital converter. The allowable analog input voltage range at each analog input pin is 0–5 V. Since each analog input pin is connected to a 10-bit analog-to-digital converter, 0–5 V is divided into 1,024 steps. So there are 1,024 steps ranging from (0 to 1,023) steps. The analogRead function returns an integer value in the range of (0–1,023). The statement (1) `int analogInput=A0` assigns name "analogInput" to pin A0 of Arduino, the analog input pin. The statement (2) `pinMode(analogInput, INPUT)` declares the "analogInput" i.e., A0 pin of Arduino, as an input pin. The statement (3) `Serial.begin(9600)` is used to initiate the serial communication between Arduino UNO board and the computer to display on the serial monitor at 9,600 baud. The statement (4) `int step=analogRead(analogInput)` is used to read the analog value of A0 pin of Arduino UNO board. Since the analog input pin is connected to a 10-bit analog-to-digital converter, (0–5) V is divided into 1,024 steps. So there are 1,024 steps ranging from (0 to 1,023) steps. The "analogRead" function returns an integer value in the range of (0–1,023) and assigns it to "step". The statements (5), (6), and (7) are used to display the step on the serial monitor at an interval of 1 second.

Program 4.28

Interface a potentiometer with Arduino UNO board, and write a program to display the various steps of internal analog-to-digital converter and it's equivalent voltage on the serial monitor when the potentiometer knob is rotated.

Solution

The interfacing of a potentiometer with Arduino UNO board is shown in Figure 4.48. A program to display the various steps of internal analog-to-digital converter and its equivalent voltage on the serial monitor when the potentiometer knob is rotated is shown in Figure 4.51. The screenshot of serial monitor of program shown in Figure 4.51 and interfacing circuit shown in Figure 4.48 to display the various steps of internal analog-to-digital converter and its equivalent voltage on the serial monitor when the potentiometer knob is rotated is shown in Figure 4.52.

Description of the Program:

The statements (1), (2), and (3) initialize the analog input pin A0 as "analogInput" and the serial communication between Arduino UNO board and the computer to display on the serial monitor at 9,600 baud.

The statement (4) `int step=analogRead(analogInput)` is used to assign 0–1,023 steps to integer "step" depending upon the position of Terminal 2 (wiper) of the potentiometer.

int analogInput=A0;	*statement (1)*
void setup()	
{	
pinMode(analogInput,INPUT);	*statement (2)*
Serial.begin(9600);	*statement (3)*
}	
void loop()	
{	
int step=analogRead(analogInput);	*statement (4)*
float voltage=step(5.0/1023.0);*	*statement (5)*
Serial.print("Step Number= ");	*statement (6)*
Serial.print(step);	*statement (7)*
Serial.print(" Equivalent Voltage= ");	*statement (8)*
Serial.println(voltage);	*statement (9)*
delay(1000);	*statement (10)*
}	

FIGURE 4.51 An Arduino UNO program to display the various steps of internal analog-to-digital converter and its equivalent voltage on the serial monitor when the potentiometer knob is rotated for the circuit diagram shown in Figure 4.48.

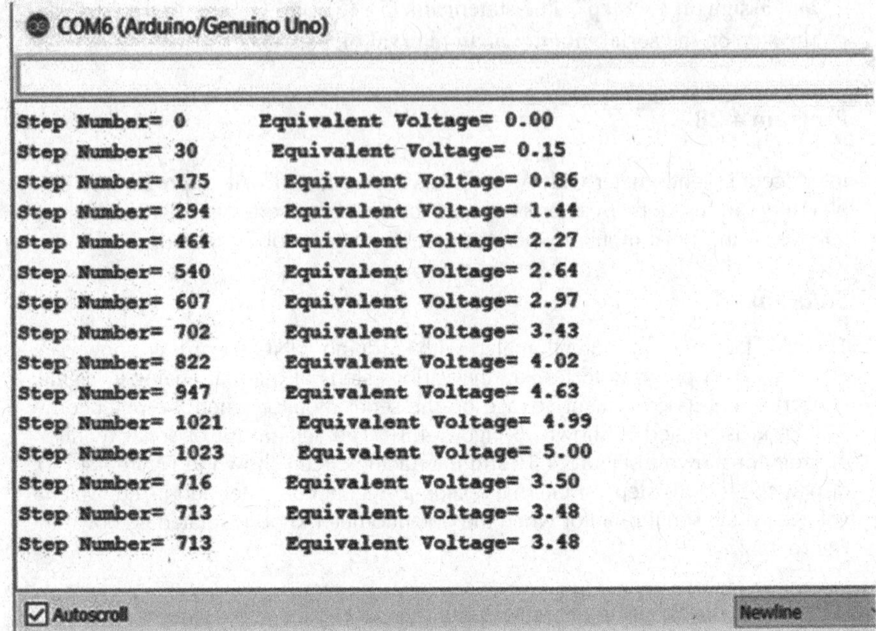

FIGURE 4.52 The screenshot of serial monitor of program shown in Figure 4.51 and interfacing circuit shown in Figure 4.48 to display the various steps of internal analog-to-digital converter and its equivalent voltage on the serial monitor when the potentiometer knob is rotated.

The statement (5) `float voltage=step*(5.0/1023.0)` will convert the step into its equivalent voltage in the range 0–5 V, where step 0 represents 0 V and step 1,023 represents 5 V.

The statements (6), (7), (8), (9), and (10) are used to display the step and its equivalent voltage on the serial monitor at an interval of 1 second.

Program 4.29

Re-write the program shown in Figure 4.51 by using "map" function.

Solution

An Arduino UNO program for the circuit diagram shown in Figure 4.48 to display on the serial monitor the various steps of internal analog-to-digital converter and its equivalent voltage using "map" function when the potentiometer knob is rotated is shown in Figure 4.53. The screenshot of serial monitor of program shown in Figure 4.53 and interfacing circuit shown in Figure 4.48 to display the various steps of internal analog-to-digital converter and its equivalent voltage using "map" function on the serial monitor when the potentiometer knob is rotated is shown in Figure 4.54.

int analogInput=A0;	*statement (1)*
void setup()	
{	
pinMode(analogInput,INPUT);	*statement (2)*
Serial.begin(9600);	*statement (3)*
}	
void loop()	
{	
int step=analogRead(analogInput);	*statement (4)*
float voltage=map(step,0,1023,0.0,5.0);	*statement (5)*
Serial.print("Step Number= ");	*statement (6)*
Serial.print(step);	*statement (7)*
Serial.print(" Equivalent Voltage= ");	*statement (8)*
Serial.println(voltage);	*statement (9)*
delay(1000);	*statement (10)*
}	

FIGURE 4.53 An Arduino UNO program to display the various steps of internal analog-to-digital converter and its equivalent voltage using "map" function on the serial monitor when the potentiometer knob is rotated for the circuit diagram shown in Figure 4.48.

COM6 (Arduino/Genuino Uno)

```
Step Number= 0    Equivalent Voltage= 0.00
Step Number= 38    Equivalent Voltage= 0.00
Step Number= 0    Equivalent Voltage= 0.00
Step Number= 24    Equivalent Voltage= 0.00
Step Number= 70    Equivalent Voltage= 0.00
Step Number= 262    Equivalent Voltage= 1.00
Step Number= 263    Equivalent Voltage= 1.00
Step Number= 424    Equivalent Voltage= 2.00
Step Number= 424    Equivalent Voltage= 2.00
Step Number= 549    Equivalent Voltage= 2.00
Step Number= 670    Equivalent Voltage= 3.00
Step Number= 764    Equivalent Voltage= 3.00
Step Number= 783    Equivalent Voltage= 3.00
Step Number= 1021    Equivalent Voltage= 4.00
Step Number= 1023    Equivalent Voltage= 5.00
```

Autoscroll   Newline

FIGURE 4.54 The screenshot of serial monitor of program shown in Figure 4.53 and interfacing circuit shown in Figure 4.48 to display the various steps of internal analog-to-digital converter and its equivalent voltage using "map" function on the serial monitor when the potentiometer knob is rotated.

Description of the Program:

The program written in this example is different from the program shown in Figure 4.51 only in the statement (5).

The statement (5) in the program shown in Figure 4.51 is `float voltage=step*(5.0/1023.0)`, whereas the statement (5) in the program shown in Figure 4.53 is `float voltage=map(step, 0,1023,0.0,5.0)`. Here, the values from (0 to 1,023) belong to "step". The numbers in the range (0–1,023) are mapped in between (0.0–5.0) V, where step value 0 mapped with 0.0 V and 1,023 mapped with 5.0 V.

Program 4.30

Interface a potentiometer and LED with Arduino UNO board, and write a program to turn on the LED when the voltage across the potentiometer is greater than or equal to 2.5 V and turn off the LED when the voltage across the potentiometer is less than 2.5 V.

Solution

The interfacing of potentiometer with Arduino UNO board is shown in Figure 4.55. The Terminals T1 and T2 of 10 KΩ potentiometer are connected to the 5 V and GND (ground) pin of Arduino board. The wiper terminal of 10 KΩ potentiometer is connected to A0 pin of Arduino board. The anode of LED is connected to Pin 13 of Arduino UNO board through a 250 Ω resistor, and the cathode is connected to the GND (ground) pin of Arduino board.

A program to turn on the LED when the voltage across the potentiometer is greater than or equal to 2.5 V and turn off the LED when the voltage across the potentiometer is less than 2.5 V is shown in Figure 4.56. The screenshot of serial monitor of program is shown in Figure 4.56 and interfacing circuit is shown in Figure 4.55 to display the voltage across the potentiometer and the status of the LED.

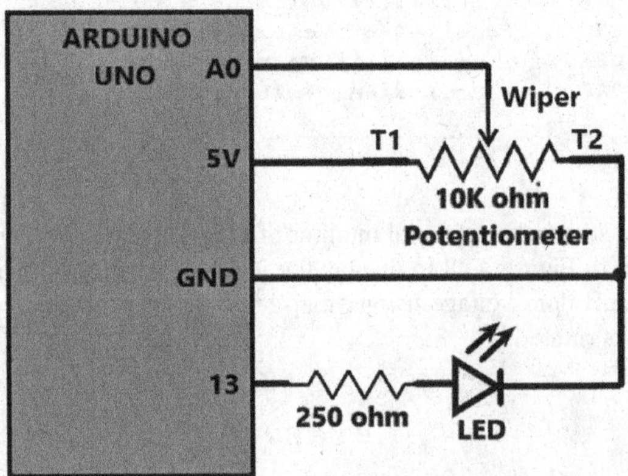

FIGURE 4.55 The interfacing of a potentiometer and LED with Arduino UNO board.

int led=13;	*statement (1)*
int analogInput=A0;	*statement (2)*
void setup()	
{	
pinMode(led,OUTPUT);	*statement (3)*
pinMode(analogInput,INPUT);	*statement (4)*
Serial.begin(9600);	*statement (5)*
}	
void loop()	
{	
int step=analogRead(analogInput);	*statement (6)*
float voltage= step(5.0/1024.0);*	*statement (7)*
if (voltage >= 2.5)	*statement (8)*
{	
digitalWrite(led,HIGH);	*statement (9)*
Serial.print("voltage= ");	*statement (10)*
Serial.println(voltage);	*statement (11)*
Serial.println("LED is ON");	*statement (12)*
}	
else	
{	
digitalWrite(led,LOW);	*statement (13)*
Serial.print("voltage= ");	*statement (14)*
Serial.println(voltage);	*statement (15)*
Serial.println("LED is OFF");	*statement (16)*
}	
delay(1000);	*statement (17)*
}	

FIGURE 4.56 An Arduino UNO program to turn on the LED when the voltage across the potentiometer is greater than or equal to 2.5 V and turn off the LED when the voltage across the potentiometer is less than 2.5 V for the circuit diagram shown in Figure 4.55.

Description of the Program:

After the execution of statements from (1) to (7), the variable "voltage" will get a value in the range from 0–5 V when the knob of the potentiometer is rotated.

When the value of "voltage" variable is greater than or equal to 2.5 V, statement (8) will be valid and statements (9) to (12) will be executed to turn on the LED and display "LED is ON" on the serial monitor; otherwise, statements (13) to (16) will be executed to turn off the LED and display "LED is OFF" on the serial monitor (Figure 4.57).

Program 4.31

Interface a potentiometer and two LEDs (red and green) with Arduino UNO board, and write a program to turn on the red LED and turn off the green LED when the voltage across the potentiometer is greater than 2.5 V and turn off the red LED and turn on the green LED when the voltage across the potentiometer is less than or equal to 2.5 V.

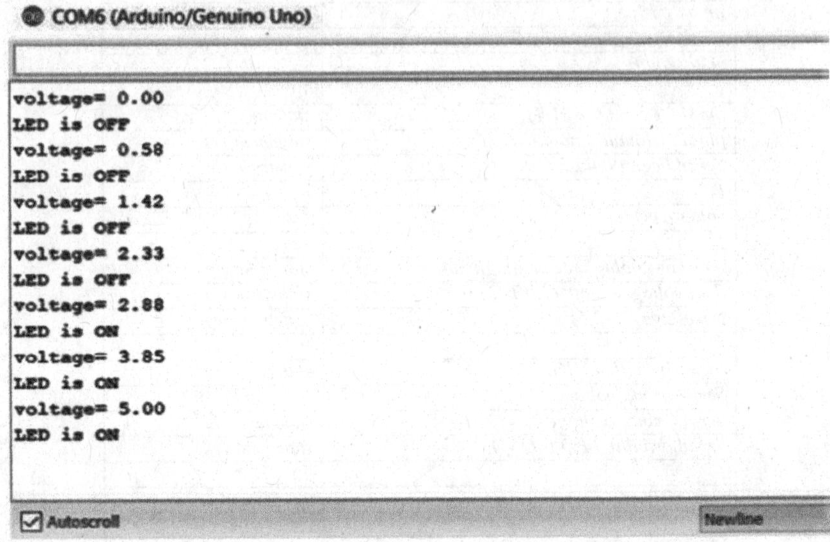

FIGURE 4.57 The screenshot of serial monitor of program shown in Figure 4.56 and interfacing circuit shown in Figure 4.55 to display the voltage across the potentiometer and the status of the LED.

Solution

The interfacing of potentiometer and LEDs with Arduino UNO board is shown in Figure 4.58. The Terminals T1 and T2 of 10 KΩ potentiometer are connected to the 5 V and GND (ground) pin of Arduino board. The wiper terminal of 10 KΩ potentiometer is connected to A0 pin of Arduino board. The anode of green LED is connected to Pin 12 of Arduino UNO board through a 250 Ω resistor, and the cathode is connected to the GND (ground) pin of Arduino board. The anode of red LED is connected to Pin 13 of Arduino UNO board through a 250 Ω resistor, and the cathode is connected to the GND (ground) pin of Arduino board.

A program to turn on the red LED and turn off the green LED when the voltage across the potentiometer is greater than 2.5 V and turn off the red LED and turn on the green LED when the voltage across the potentiometer is less than or equal to 2.5 V is shown in Figure 4.59.

The screenshot of serial monitor of program shown in Figure 4.59 and interfacing circuit shown in Figure 4.58 to display the voltage across the potentiometer and the status of the red and green LED is shown in Figure 4.60.

Description of the Program:

The statements (1), (2) and (3) initialize Pin 12 to `greenLED`, Pin 13 to `redLED`, and the analog input pin A0 as `analogInput`. The statements (4) and (5) will initialize digital Pins 12 and 13 as output pins, and the statement (6) will initialize the analog pin A0 as an input pin. The statement (7) initializes serial communication between the Arduino UNO board and the computer to display the serial monitor at 9,600 baud.

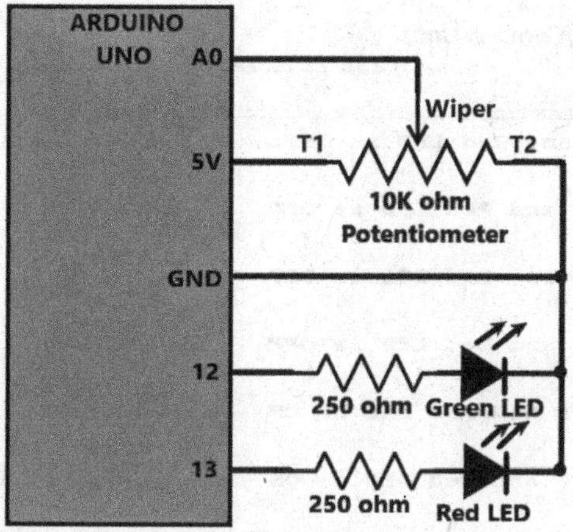

FIGURE 4.58 The interfacing of a potentiometer and LED with Arduino UNO board.

int greenLED=12;	*statement (1)*
int redLED=13;	*statement (2)*
int analogInput=A0;	*statement (3)*
void setup()	
{	
pinMode(greenLED,OUTPUT);	*statement (4)*
pinMode(redLED,OUTPUT);	*statement (5)*
pinMode(analogInput,INPUT);	*statement (6)*
Serial.begin(9600);	*statement (7)*
}	
void loop()	
{	
int step=analogRead(analogInput);	*statement (8)*
float voltage= step(5.0/1023.0);*	*statement (9)*
if (voltage <= 2.5)	*statement (10)*
{	
digitalWrite(greenLED,HIGH);	*statement (11)*
digitalWrite(redLED,LOW);	*statement (12)*
Serial.print("voltage= ");	*statement (13)*
Serial.println(voltage);	*statement (14)*
Serial.println("Green LED is ON and Red LED is OFF");	*statement (15)*
}	
else if (voltage > 2.5)	*statement (16)*
{	
digitalWrite(greenLED,LOW);	*statement (17)*
digitalWrite(redLED,HIGH);	*statement (18)*
Serial.print("voltage= ");	*statement (19)*
Serial.println(voltage);	*statement (20)*
Serial.println("Green LED is OFF and Red LED is ON");	*statement (21)*
}	
delay(1000);	*statement (22)*
}	

FIGURE 4.59 An Arduino UNO program to turn on the red LED and turn off the green LED when the voltage across the potentiometer is greater than 2.5 V and turn off the red LED and turn on the green LED when the voltage across the potentiometer is less than or equal to 2.5 V for the circuit diagram shown in Figure 4.58.

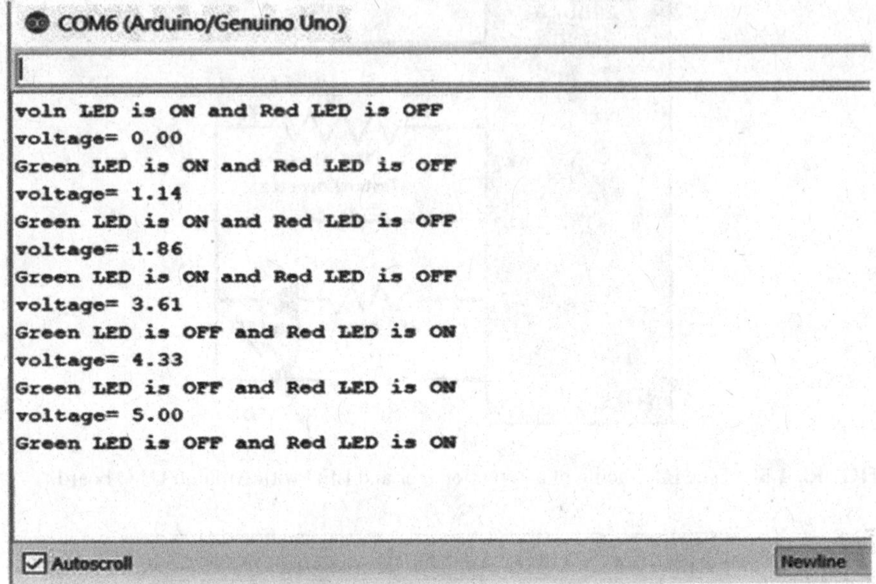

FIGURE 4.60 The screenshot of serial monitor of program shown in Figure 4.59 and interfacing circuit shown in Figure 4.58 to display the voltage across the potentiometer and the status of the red and green LED.

The statement (8) `int step=analogRead(analogInput)` is used to assign 0–1,023 steps to integer "step" depending upon the position of Terminal 2 (wiper) of the potentiometer.

The statement (9) `float voltage=step*(5.0/1023.0)` will convert the step into its equivalent voltage in the range 0–5 V, where step 0 represents 0 V and step 1,023 represents 5 V.

When the value of "voltage" variable is less than or equal to 2.5 V, statement (10) will be true and statements (11) to (15) will be executed to turn on the green LED and turn off red LED and display "Green LED is ON and Red LED is OFF" on the serial monitor.

When the value of "voltage" variable is greater than 2.5 V, statement (16) will be true and statements (17) to (21) will be executed to turn off the green LED and turn on red LED and display "Green LED is OFF and Red LED is ON" on the serial monitor.

The above process will be repeated after every second due to statement (22) `delay(1000)`.

4.8 ARDUINO PROGRAMMING USING PWM TECHNIQUES

This section shall discuss some programming concepts related to PWM concepts and its implementation by using the Arduino UNO board. The PWM is a technique by which we can indirectly encode the digital value into an equivalent analog value.

There are six pins in the Arduino UNO board, namely, 3, 5, 6, 9, 10, and 11, with PWM capabilities. The six PWM pins are labeled as "~". The concept of PWM is explained in Section 3.7 of Chapter 3.

Program 4.32

Interface a LED and two push-button switches with Arduino UNO board, and write a program to turn on the LED with low intensity when Switch 1 is pressed and turn on it with high intensity when Switch 2 is pressed.

Solution

The interfacing of two push-button switches and one LED with Arduino UNO board is shown in Figure 4.61. The Terminal T2 of Switch 1 and Switch 2 is connected to the GND (ground) pin of Arduino board, and the Terminal T1 of the switches is connected to the one terminal of two 1 KΩ resistors. The other terminals of 1 KΩ resistor are connected to the 5 V pin of Arduino board. The junction of Terminal T1 of Switch 1 and 1 KΩ resistor is extended and connected to Pin 2 of Arduino board. The junction of Terminal T1 of Switch 2 and 1 KΩ resistor is extended and connected to Pin 3 of Arduino board. The anode of the LED is connected to Pin 11 (PWM output pin) of Arduino UNO board through a 250 Ω resistor, and the cathode is connected to the GND (ground) pin of Arduino board.

An Arduino UNO program to turn on the LED with low intensity when Switch 1 is pressed and turn on the LED with high intensity when Switch 2 is pressed is shown in Figure 4.62. The screenshot of serial monitor of program shown in Figure 4.62 and interfacing circuit shown in Figure 4.61 to display the status of the LED is shown in Figure 4.63.

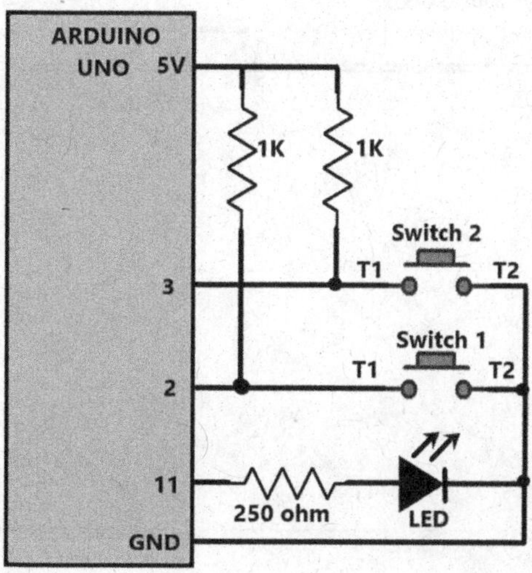

FIGURE 4.61 The interfacing of a LED and two push-button switches with Arduino UNO board.

int pushbuttonSwitch1 = 2;	statement (1)
int pushbuttonSwitch2 = 3;	statement (2)
int LED = 11;	statement (3)
void setup()	
{	
pinMode(pushbuttonSwitch1,INPUT);	statement (4)
pinMode(pushbuttonSwitch2,INPUT);	statement (5)
pinMode(LED,OUTPUT);	statement (6)
Serial.begin(9600);	statement (7)
}	
void loop()	
{	
int buttonState1 =digitalRead(pushbuttonSwitch1);	statement (8)
int buttonState2 =digitalRead(pushbuttonSwitch2);	statement (9)
if (buttonState1==0)	statement (10)
{	
analogWrite(LED,100);	statement (11)
Serial.println("LED ON 'LOW'");	statement (12)
}	
else if (buttonState2==0)	statement (13)
{	
analogWrite(LED,250);	statement (14)
Serial.println("LED ON 'HIGH'");	statement (15)
}	
delay(1000);	statement (16)
}	

FIGURE 4.62 An Arduino UNO program to turn on the LED with low intensity when Switch 1 is pressed and turn on the LED with high intensity when Switch 2 is pressed for the circuit diagram shown in Figure 4.61.

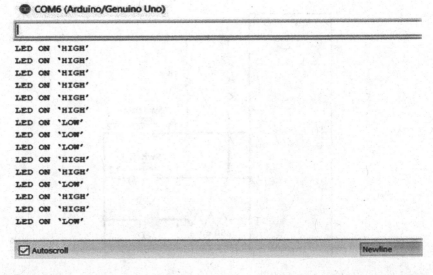

FIGURE 4.63 The screenshot of serial monitor of program shown in Figure 4.62 and interfacing circuit shown in Figure 4.61 to display the status of the LED.

Description of the Program:

The statements (1) and (2) initialize Pins 2 and 3 of Arduino UNO board to connect two push-button switches Switch 1 and Switch 2.

The statement (3) initializes Pin 11 to connect a LED. Here remember that in this program, we shall use the pulse width modulation capability of Pin 11. The Arduino UNO board contains six pins, namely, 3, 5, 6, 9, 10, and 11, with PWM capabilities. The six PWM pins are labeled as "~".

The statements (4) and (5) initialize digital Pins 2 and 3 as input pins, and the statement (6) initializes Pin 11 as an output pin. The statement (7) initializes serial communication between the Arduino UNO board and the computer to display the serial monitor at 9,600 baud.

The statements (8) and (9) assign the digital value of Switch 1 and Switch 2 connected at Pins 2 and 3, respectively, to the variables "buttonState1" and "buttonState2".

Refer to Figure 4.61, it can be understood that whenever Switch 1 or Switch 2 is pressed, then Pin 2 or 3 will get 0 values, and whenever Switch 1 or Switch 2 is not pressed, then Pin 2 or 3 will get 1 value.

Due to the statement (10) if Switch 1 is pressed, then statements (11) and (12) will be executed. In Section 3.7, we have already discussed that a value 255 generates 100% duty cycle wave from a PWM pin of Arduino board; therefore, the 100 value in the statement (11) `analogWrite(LED, 100)`, will generate 39.2% duty cycle wave. We also know that 100% duty cycle wave generates 5 V analog signal from PWM pin; therefore, 39.2% duty cycle wave generates 1.96 V analog signal from Pin 11, and thus, LED will turn on with lower intensity.

Due to the statement (13) if Switch 2 is pressed, then statements (14) and (15) will be executed. Due to the statement (14) `analogWrite(LED, 250)`, the 250 value will generate a 98.0% duty cycle wave, which generates 4.9 V analog signal from Pin 11, and thus, LED will turn on with higher intensity.

Program 4.33

Interface a LED and a potentiometer with Arduino UNO board, and write a program to vary the turn-on intensity of LED by rotating the knob of the potentiometer.

Solution

The interfacing of a 10 KΩ potentiometer and LED with Arduino UNO board is shown in Figure 4.64. The Terminals T1 and T2 of 10 KΩ potentiometer are connected to the 5 V and GND (ground) pin of Arduino board. The wiper terminal of 10 KΩ potentiometer is connected to A0 pin of Arduino board. The anode of the LED is connected to Pin 9 (PWM output pin) of Arduino UNO board through a 250 Ω resistor, and the cathode is connected to the GND (ground) pin of Arduino board.

An Arduino UNO program to vary the turn-on intensity of LED by rotating the knob of the potentiometer for the circuit diagram shown in Figure 4.64 is shown in Figure 4.65.

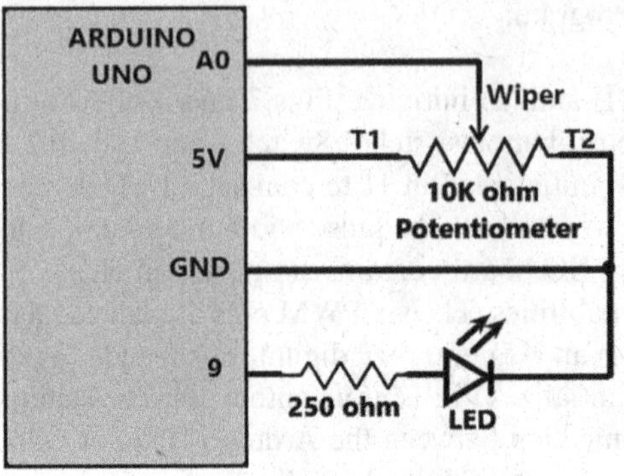

FIGURE 4.64 The interfacing of a LED and a potentiometer with Arduino UNO board.

int analogInput= A0;	*statement (1)*
int LED = 9;	*statement (2)*
void setup()	
{	
pinMode(analogInput,INPUT);	*statement (3)*
pinMode(LED,OUTPUT);	*statement (4)*
Serial.begin(9600);	*statement (5)*
}	
void loop()	
{	
int step=analogRead(analogInput);	*statement (6)*
float pwmOutput=step(255.0/1023.0);*	*statement (7)*
float dutyCycle=pwmOutput(100.0/255.0);*	*statement (8)*
float voltage= step(5.0/1023.0);*	*statement (9)*
analogWrite(LED,pwmOutput);	*statement (10)*
Serial.print("step= ");	*statement (11)*
Serial.print(step);	*statement (12)*
Serial.print(" PWM Output = ");	*statement (13)*
Serial.print(pwmOutput);	*statement (14)*
Serial.print(" Duty Cycle (%) = ");	*statement (15)*
Serial.print(dutyCycle);	*statement (16)*
Serial.print(" Analog Voltage= ");	*statement (17)*
Serial.println(voltage);	*statement (18)*
delay(1000);	*statement (19)*
}	

FIGURE 4.65 An Arduino UNO program to vary the turn-on intensity of LED by rotating the knob of the potentiometer for the circuit diagram shown in Figure 4.64.

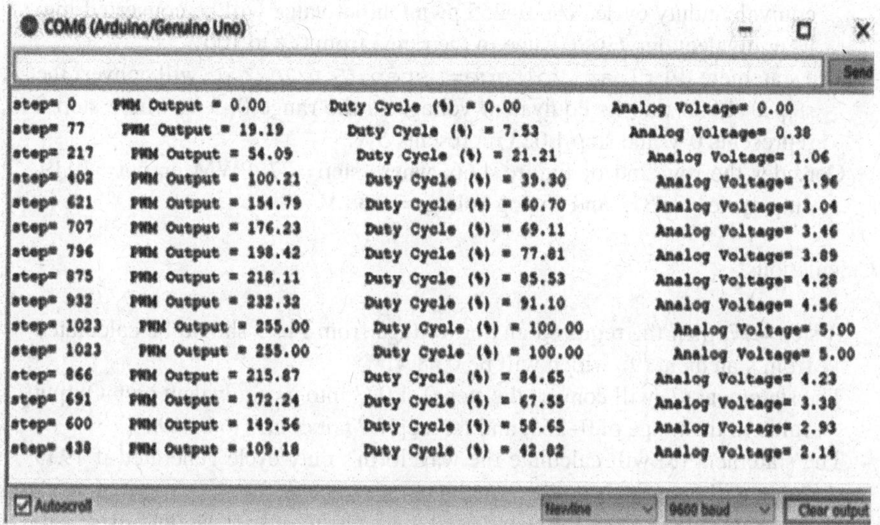

FIGURE 4.66 The screenshot of serial monitor of program shown in Figure 4.65 and interfacing circuit shown in Figure 4.64 to display the steps of internal analog-to-digital converter, PWM output from Pin 9 of Arduino board, duty cycle, and the voltage across the potentiometer.

The screenshot of serial monitor of program shown in Figure 4.65 and interfacing circuit shown in Figure 4.64 to display the steps of internal analog-to-digital converter, PWM output from Pin 9 of Arduino board, duty cycle, and the voltage across the potentiometer is shown in Figure 4.66.

Description of the Program:

The statements (1) and (2) initialize Pins A0 and 9 of Arduino UNO board to connect analog input and LED. In this program, we shall use the pulse width modulation capability of Pin 9.

The statement (3) initializes A0 pin as an input pin, and the statement (4) initializes Pin 9 as an output pin.

The statement (5) initializes serial communication between the Arduino UNO board and the computer to display the serial monitor at 9,600 baud.

The statement (6) `int step=analogRead(analogInput)` is used to assign 0–1,023 steps to integer “step” depending upon the position of Terminal 2 (wiper) of the potentiometer.

The statement (7) `float pwmOutput=step*(255.0/1023.0)` will convert the step 0–1,023 into its equivalent pwmOutput value in the range from 0 to 255. These pwmOutput values are required for the generation of the required duty cycle waveform.

The statement (8) `float dutyCycle=pwmOutput*(100.0/255.0)` will convert the “pwmOutput” value generated in statement (7) into its equivalent duty cycle. The 0–255 pwmOutput value will be converted into its equivalent dutyCycle value in the range from 0% to 100%.

The statement (9) `float voltage= step*(5.0/1023.0)` will convert the step 0–1,023 into its equivalent voltage in the range 0–5 V, where step 0 represents 0 V and step 1,023 represents 5 V.

Consider the row 2nd of Figure 4.66, where step=77, PWM output=19.19, duty cycle=7.53%, and analog voltage=0.38 V.

Calculations -

If step=77, then the required analog voltage from Pin 9 should be calculated from statement (9), which will be 0.38 V.

The statement (7) will convert the step 0–1,023 into its equivalent pwmOutput value in the range of 0–255, and for step=77, it comes to be 19.19.

The statement (8) will calculate the waveform's duty cycle generated if 19.19 value will be written in Pin 9, and it comes to be 7.53%.

To generate 0.38 V from Pin 9, we need a waveform of specific duty cycle calculated from statement (15), and it comes to be 19.19.

Program 4.34

Interface a LED with Arduino UNO board, and write a program to vary its intensity by using PWM technique.

Solution

The interfacing of a LED with Arduino UNO board is shown in Figure 4.67. The anode of the LED is connected to Pin 9 (PWM output pin) of Arduino UNO board through a 250 Ω resistor, and the cathode is connected to the GND (ground) pin of Arduino board.

An Arduino UNO program to vary the on intensity of LED by using PWM technique for the circuit diagram shown in Figure 4.67 is shown in Figure 4.68.

Description of the Program:

The statement (1) initializes Pin 9 of the Arduino UNO board to connect LED. In this program, we shall use the pulse width modulation capability of Pin 9.

The statement (2) and statement (3) declare integer-type brightness and brightnessVariation with initializing values 0 and 5, respectively.

The statement (4) initializes Pin 9, where we will connect LED as an output pin.

The statement (5) `analogWrite(led, brightness)` will write the value of "brightness", which is 0 initially, to Pin 9, and by doing so, Pin 9 will generate 0 V, which will cause LED connected to Pin 9 off.

The statement (6) `brightness=brightness + brightnessVariation` will make brightness=5 since brightnessVariation=5.

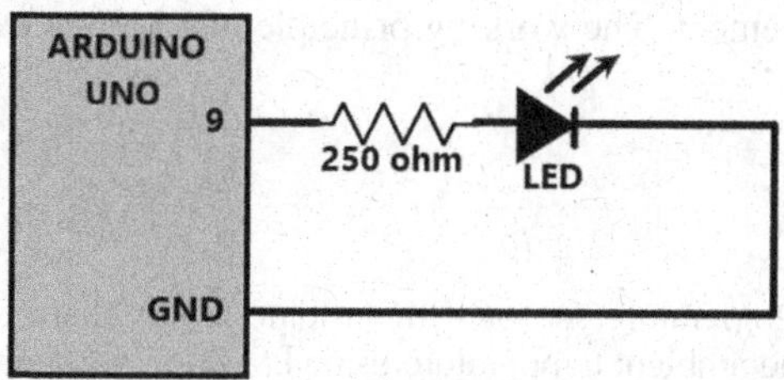

FIGURE 4.67 The interfacing of a LED with Arduino UNO board for Program 4.34.

int led=9;	*statement (1)*
int brightness=0;	*statement (2)*
int brightnessVariation=5;	*statement (3)*
void setup()	
{	
pinMode(led,OUTPUT);	*statement (4)*
}	
void loop()	
{	
analogWrite(led,brightness);	*statement (5)*
brightness=brightness + brightnessVariation;	*statement (6)*
if (brightness==0 || brightness==255)	*statement (7)*
{	
brightnessVariation = - brightnessVariation;	*statement (8)*
}	
delay(30);	*statement (9)*
}	

FIGURE 4.68 An Arduino UNO program to vary the on intensity of LED by using PWM technique for the circuit diagram shown in Figure 4.67.

The statement (7) `if (brightness==0 || brightness==255)` will generate "True" only if brightness==0 or brightness==255, then only statement (8) will be executed; otherwise, statement (5) and statement (6) execute one after other continuously. Due to the continuous execution of statement (5) and statement (6), the brightness of LED increases in the step of 5. The LED will turn on with full intensity when brightness = 255. Once the value of brightness equals 255, the statement (7) becomes true and statement (8) `brightnessVariation = - brightnessVariation` will be executed. Now the value of brightness decreases with the step of 5, and this will cause LED to turn on with reducing intensity.

4.9 INTERFACING AND PROGRAMMING OF ARDUINO WITH LM35

This section shall discuss the interfacing of temperature sensor with Arduino UNO board and programming details. The LM35 temperature sensor is used for the temperature measurement. The working principle of LM35 is explained in Section 3.8 of Chapter 3.

Program 4.35

Interface LM35 temperature sensor with Arduino UNO board, and write a program to display the ambient temperature using LM35 on the serial monitor.

Solution

The interfacing of LM35 temperature sensor with Arduino UNO board is shown in Figure 4.69. The Pins 1, 2, and 3 of LM35 are connected to the 5 V, A0, and GND (ground) pins of Arduino board. The analog voltage proportional to the external ambient temperature will be generated from Pin 2 of LM35 and connected to the Arduino UNO board's analog input pin A0.

An Arduino UNO program to display the ambient temperature using LM35 on the serial monitor for the circuit diagram shown in Figure 4.69 is shown in Figure 4.70.

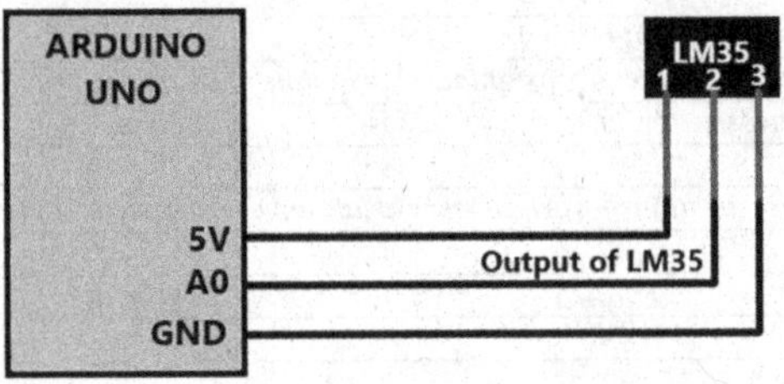

FIGURE 4.69 The interfacing of LM35 with Arduino UNO board.

int analogVoltage=A0;	*statement (1)*
void setup()	
{	
pinMode(analogVoltage,INPUT);	*statement (2)*
Serial.begin(9600);	*statement (3)*
}	
void loop()	
{	
int step=analogRead(A0);	*statement (4)*
*float temp= step*0.48828125;*	*statement (5)*
Serial.print("Room Temperature= ");	*statement (6)*
Serial.print(temp);	*statement (7)*
Serial.println("°C");	*statement (8)*
delay(2000);	*statement (9)*
}	

FIGURE 4.70 An Arduino UNO program to display the ambient temperature using LM35 on the serial monitor for the circuit diagram shown in Figure 4.69.

The screenshot of serial monitor of program shown in Figure 4.70 and interfacing circuit shown in Figure 4.69 to display the ambient temperature using LM35 on the serial monitor is shown in Figure 4.71 (a).

The screenshot of serial monitor of program shown in Figure 4.70 and interfacing circuit shown in Figure 4.69 to display the temperature when we touch LM35 with our finger is shown in Figure 4.71 (b).

Calculation of multiplying factor 0.48828125 in statement (5) "float temp= step*0.48828125" – The analog voltage proportional to the external ambient temperature will be generated from Pin 2 of LM35 and connected to the Arduino UNO board's analog input pin A0.

Internally A0 pin is connected to a 10-bit analog-to-digital converter. The internal ADC of the Arduino UNO board has 1,024 steps ranging from 0 to 1,023. The allowable analog input voltage range at A0 is 0–5 V and divided into 1,024 steps with a step size of 4.88281 mV [Calculated from equation (3.5)].

Since LM35 generates 10 mV per °C rise in temperature, 1 mV represents 0.1°C, and 4.88281 mV (one step size) represents (0.1°C×4.88281), and it comes to be 0.488281°C. Therefore, to find the temperature corresponding to the analog signal generated by the LM35 temperature sensor, we must multiply the analogue signal step with 0.488281°C.

Description of the Program:

The statements (1), (2), and (3) initialize the analog input pin A0 as "analogInput" and the serial communication between Arduino UNO board and the computer to display on the serial monitor at 9,600 baud.

The statement (4) `int step=analogRead(analogInput)` is used to assign 0–1,023 steps to integer "step" depending upon the analog voltage generated in response to the ambient temperature.

The statement (5) `float temp= step*0.48828125` will convert the step into its equivalent temperature in degree centigrade.

The statements (6) to (9) are used to display the ambient temperature on the serial monitor with the gap of 2 seconds.

The snapshot of displaying the ambient temperature on the serial monitor is shown in Figure 4.71(a). If we hold the LM35 sensor with our fingers for 5–6 seconds, our body heats the LM35, and LM35 will sense the raised temperature. The snapshot of displaying the temperature on the serial monitor when we hold the LM35 sensor with our fingers is shown in Figure 4.71(b).

Program 4.36

Interface an LM35 temperature sensor and LED with Arduino UNO board, and write a program to turn on the LED if the room temperature is greater than or equal to 30.0°C and turn off the LED if the room temperature is less than 30.0°C. Display the room temperature and status of LED on the serial monitor.

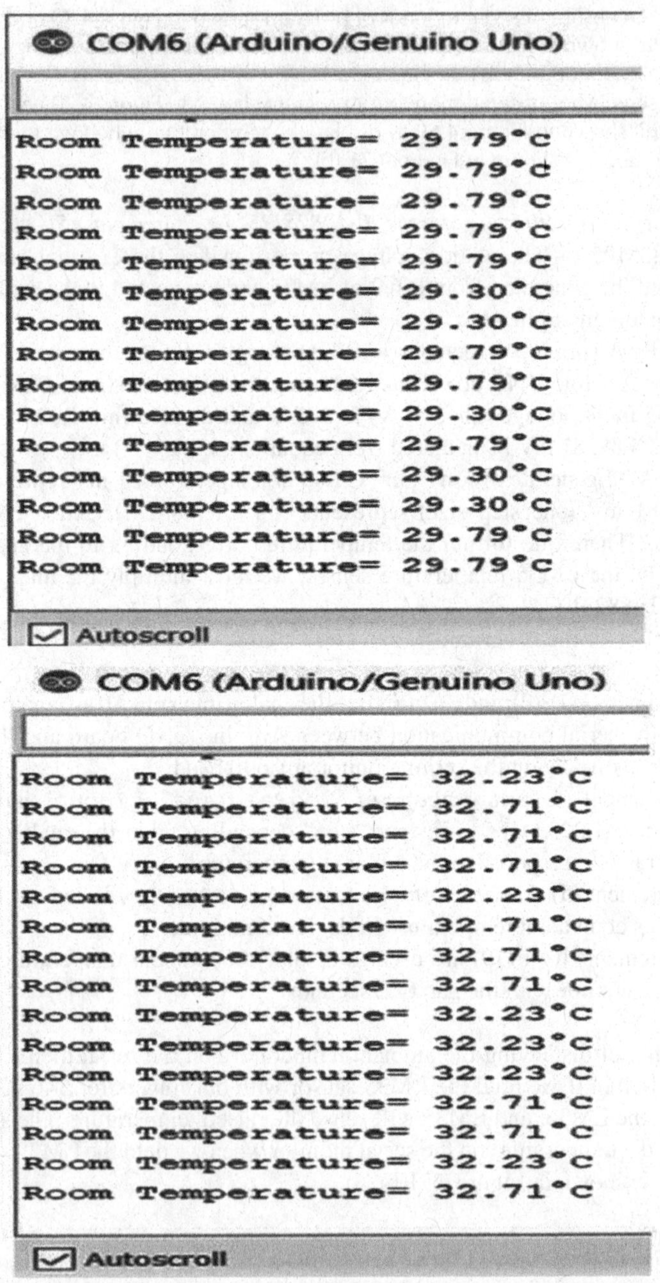

FIGURE 4.71 (a) The screenshot of serial monitor of program shown in Figure 4.70 and interfacing circuit shown in Figure 4.69 to display the ambient temperature using LM35 on the serial monitor. (b) The screenshot of serial monitor of program shown in Figure 4.70 and interfacing circuit shown in Figure 4.69 to display the temperature when we touch LM35 with our finger.

Solution

The interfacing of LM35 temperature sensor and LED with Arduino UNO board is shown in Figure 4.72. The Pins 1, 2, and 3 of LM35 are connected to the 5 V, A0, and GND (ground) pins of Arduino board. The anode of the LED is connected to Pin 9 of Arduino UNO board through a 250 Ω resistor, and the cathode is connected to the GND (ground) pin of Arduino board.

An Arduino UNO program to turn on the LED if the room temperature is greater than or equal to 30.0°C and turn off the LED if the room temperature is less than 30.0°C for the circuit diagram shown in Figure 4.72 is shown in Figure 4.73.

The screenshot of serial monitor of program shown in Figure 4.73 and interfacing circuit shown in Figure 4.72 to display the ambient temperature and the status of LED is shown in Figure 4.74.

Description of the Program:

The statements (1) and (2) initialize the analog input pin A0 as "analogInput" and Pin 9 as LED. The statements (3) and (4) initialize the A0 pin as an input pin and Pin 9 as an output pin. The statement (5) initializes the serial communication between the Arduino UNO board and the computer to display the serial monitor at 9,600 baud. The statement (6) `int step=analogRead(-analogInput)` is used to assign 0–1,023 steps to integer "step" depending upon the analog voltage generated in response to the ambient temperature. The statement (7) `float temp= step*0.48828125` will convert the step into its equivalent temperature in degree centigrade. If the condition of the statement (8) `if(temp>=30.0)` is true, then the statements (9) to (13) will be executed and LED connected at Pin 9 of Arduino board will turn on and room temperature will be displayed on the serial monitor. If the room temperature is smaller than 30°C, then statement (14) will be executed, and then statements (15) to (19) will be executed and LED connected at Pin 9 of Arduino board will turn off. Room temperature will be displayed on the serial monitor. The monitoring of temperature and corresponding operation will be repeated after every 2 seconds.

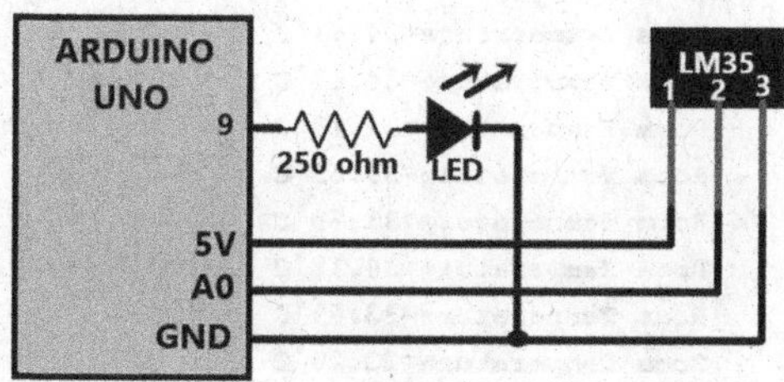

FIGURE 4.72 The interfacing of LM35 and LED with Arduino UNO board.

int analogInput=A0;	*statement (1)*
int LED=9;	*statement (2)*
void setup()	
{	
pinMode(analogInput,INPUT);	*statement (3)*
pinMode(LED,OUTPUT);	*statement (4)*
Serial.begin(9600);	*statement (5)*
}	
void loop()	
{	
int step=analogRead(A0);	*statement (6)*
*float temp= step*0.48828125;*	*statement (7)*
if(temp>=30.0)	*statement (8)*
{	
digitalWrite(LED,HIGH);	*statement (9)*
Serial.print("LED ON");	*statement (10)*
Serial.print(" Room Temperature=");	*statement (11)*
Serial.print(temp);	*statement (12)*
Serial.println("°C");	*statement (13)*
}	
else	*statement (14)*
{	
digitalWrite(LED,LOW);	*statement (15)*
Serial.print("LED OFF");	*statement (16)*
Serial.print(" Room Temperature=");	*statement (17)*
Serial.print(temp);	*statement (18)*
Serial.println("°C");	*statement (19)*
}	
delay(2000);	*statement (20)*
}	

FIGURE 4.73 An Arduino UNO program to turn on the LED if the room temperature is greater than or equal to 30.0°C and turn off the LED if the room temperature is less than 30.0°C for the circuit diagram shown in Figure 4.72.

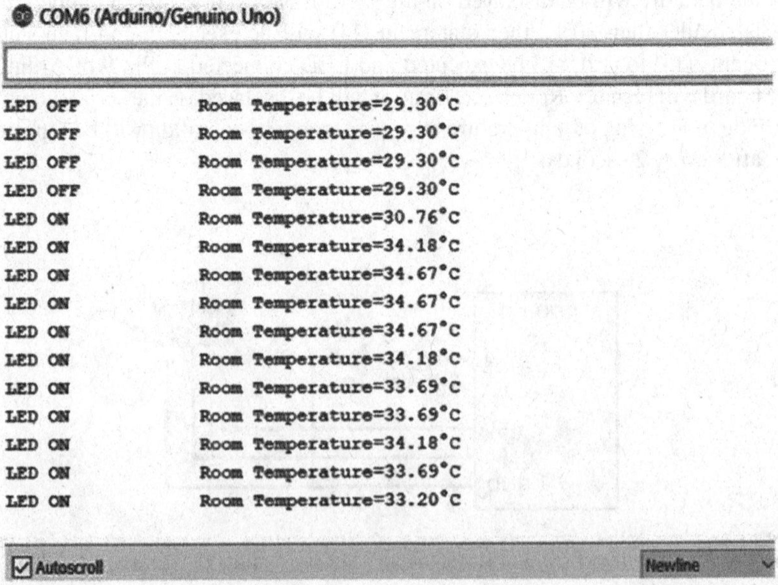

```
COM6 (Arduino/Genuino Uno)

LED OFF          Room Temperature=29.30°C
LED OFF          Room Temperature=29.30°C
LED OFF          Room Temperature=29.30°C
LED OFF          Room Temperature=29.30°C
LED ON           Room Temperature=30.76°C
LED ON           Room Temperature=34.18°C
LED ON           Room Temperature=34.67°C
LED ON           Room Temperature=34.67°C
LED ON           Room Temperature=34.67°C
LED ON           Room Temperature=34.18°C
LED ON           Room Temperature=33.69°C
LED ON           Room Temperature=33.69°C
LED ON           Room Temperature=34.18°C
LED ON           Room Temperature=33.69°C
LED ON           Room Temperature=33.20°C

[✓] Autoscroll                                   Newline
```

FIGURE 4.74 The screenshot of serial monitor of program shown in Figure 4.73 and interfacing circuit shown in Figure 4.72 to display the ambient temperature and the status of LED.

4.10 INTERFACING AND PROGRAMMING OF ARDUINO WITH HUMIDITY AND TEMPERATURE SENSOR DHT11

This section shall discuss interfacing of humidity and temperature with Arduino UNO board and programming details. The DHT11 humidity and temperature is used for the description of interfacing and programming. The working principle of DHT11 is explained in Section 3.9 of Chapter 3.

Program 4.37

Interface a DHT11 humidity and temperature sensor with Arduino UNO board, and write a program to display the serial monitor's humidity and ambient temperature.

Solution

The interfacing of DHT11 temperature and humidity sensor with Arduino UNO board is shown in Figure 4.75. The Pins 1, 2, and 4 of DHT11 are connected to the 5 V, A0, and GND (ground) pins of Arduino board. The Pin 2 of DHT11 sensor sends out the measured humidity and temperature value, and it is connected to the analog input A0 pin of the Arduino UNO board.

An Arduino UNO program to display the value of humidity and ambient temperature using DHT11 on the serial monitor for the circuit diagram shown in Figure 4.75 is shown in Figure 4.76.

The screenshot of serial monitor of program shown in Figure 4.76 and interfacing circuit shown in Figure 4.75 to display the value of humidity and ambient temperature using DHT11 on the serial monitor is shown in Figure 4.77.

Description of the Program:

The statement (1) `#include <dht.h>` includes the dht library. The dht library has all the functions required to get the humidity and temperature readings from the sensor. The dht library is not the part of in-built Arduino libraries; rather, we have to include it. For including dht library, we download the DHTLLib.zip file and copy the DHT library in Arduino Library folder.

The statement (2) `dht DHT` creates an object of name DHT. We can create an object of any name. The statements (3) and (4) initialize the analog input pin A0 as "analogInput", which is initialized as an input pin.

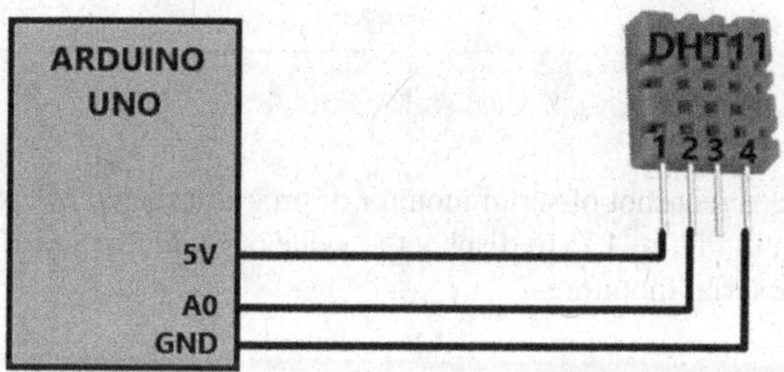

FIGURE 4.75 The interfacing of DHT11 with Arduino UNO board.

#include <dht.h>	statement (1)
dht DHT;	statement (2)
int analogInput=A0;	statement (3)
void setup()	
{	
pinMode(analogInput,INPUT);	statement (4)
Serial.begin(9600);	statement (5)
}	
void loop()	
{	
DHT.read11(A0);	statement (6)
Serial.print("Humidity= ");	statement (7)
Serial.print(DHT.humidity);	statement (8)
Serial.println("%");	statement (9)
Serial.print("temperature= ");	statement (10)
Serial.print(DHT.temperature);	statement (11)
Serial.println("°C ");	statement (12)
delay(2000);	statement (13)
}	

FIGURE 4.76 An Arduino UNO program to display the value of humidity and ambient temperature using DHT11 on the serial monitor for the circuit diagram shown in Figure 4.75.

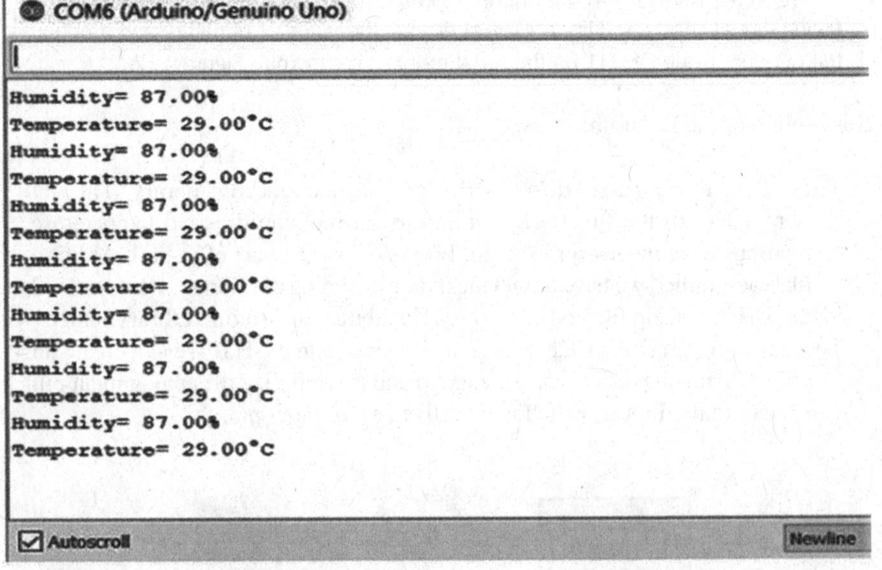

FIGURE 4.77 The screenshot of serial monitor of program shown in Figure 4.76 and interfacing circuit shown in Figure 4.75 to display the value of humidity and ambient temperature using DHT11 on the serial monitor.

The statement (5) initializes the serial communication between the Arduino UNO board and the computer to display the serial monitor at 9,600 baud. The statement (6) `DHT.read11(A0)` reads the value of humidity and temperature from analog pin A0 and assigns its value to object DHT. The humidity value can be accessed by `DHT.humidity` function, and the temperature value can be accessed by `DHT.temperature` function.

The statements from (6) to (13) are used to display the humidity value in % and temperature value in °C in serial monitor after every 2 seconds. The snapshot of the display of the humidity and ambient temperature on the serial monitor is shown in Figure 4.77.

Program 4.38

Rewrite the program shown in Figure 4.76.

Solution

The interfacing of DHT 11 humidity and temperature sensor with Arduino UNO board is shown in Figure 4.75.

A modified Arduino UNO program to display the value of humidity and ambient temperature using DHT11 on the serial monitor for the circuit diagram shown in Figure 4.75 is shown in Figure 4.78.

The screenshot of serial monitor of program shown in Figure 4.78 and interfacing circuit shown in Figure 4.75 to display the value of humidity and ambient temperature using DHT11 on the serial monitor is shown in Figure 4.79.

Description of the Program:

The statements (7) and (8) are different statements included in this program compared to the program shown in Figure 4.76. In the statement (7) `float h=DHT.humidity`, the humidity reading will be assigned to a float variable h, and in the statement (8) `float t=DHT.temperature`, the reading of temperature will be assigned to a float variable t. The statements (10) and (11) `Serial.print(h)` and `Serial.print(t)` print the value of humidity and temperature on the serial monitor. The final result of programs shown in Figures 4.76 and 4.78 is the same as shown in Figures 4.77 and 4.79.

Program 4.39

Interface a humidity and room temperature sensor DHT11 and LED with Arduino UNO, and write a program to turn on the LED and display of ambient humidity and temperature value on the serial monitor if the humidity and temperature are greater than 90% and 25°C, respectively.

#include <dht.h>	statement (1)
int analogInput=A0;	statement (2)
dht DHT;	statement (3)
void setup()	
{	
pinMode(analogInput,INPUT);	statement (4)
Serial.begin(9600);	statement (5)
}	
void loop()	
{	
DHT.read11(A0);	statement (6)
float h=DHT.humidity;	statement (7)
float t=DHT.temperature;	statement (8)
Serial.print("Humidity= ");	statement (9)
Serial.print(h);	statement (10)
Serial.println("%");	statement (11)
Serial.print("temperature= ");	statement (12)
Serial.print(t);	statement (13)
Serial.println("°C ");	statement (10)
delay(2000);	statement (11)
}	

FIGURE 4.78 A modified Arduino UNO program to display the value of humidity and ambient temperature using DHT11 on the serial monitor for the circuit diagram shown in Figure 4.75.

COM6 (Arduino/Genuino Uno)

```
Humidity= 87.00%
temperature= 29.00°C
Humidity= 87.00%
temperature= 29.00°C
Humidity= 87.00%
temperature= 29.00°C
Humidity= 87.00%
temperature= 29.00°C
Humidity= 87.00%
temperature= 29.00°C
Humidity= 87.00%
temperature= 29.00°C
Humidity= 87.00%
temperature= 29.00°C
```

FIGURE 4.79 The screenshot of serial monitor of program shown in Figure 4.78 and interfacing circuit shown in Figure 4.75 to display the value of humidity and ambient temperature using DHT11 on the serial monitor.

Solution

The interfacing of DHT11 temperature and humidity sensor and LED with Arduino UNO board is shown in Figure 4.80. The Pins 1, 2, and 4 of DHT11 are connected to the 5 V, A0, and GND (ground) pins of Arduino board. The anode of the LED is connected to Pin 8 of Arduino UNO board through a 250 Ω resistor, and the cathode is connected to the GND (ground) pin of Arduino board.

An Arduino UNO program to turn on the LED and display of ambient humidity and temperature value on the serial monitor if the humidity and temperature are greater than 90% and 25°C for the circuit diagram shown in Figure 4.80 is shown in Figure 4.81.

The screenshot of serial monitor of program shown in Figure 4.81 and interfacing circuit shown in Figure 4.80 to display the value of humidity and ambient temperature and status of LED using DHT11 on the serial monitor is shown in Figure 4.82.

Description of the Program:

The statements (1), (2), and (3) include the dht library, create an object of name DHT, and initialize the analog input pin A0 as "analogInput", respectively. The statement (4) initializes the digital I/O Pin 8 of Arduino UNO board as "LED".

The statements (5) and (6) initialize Pins A0 and 8 of Arduino UNO board as input and output.

The statement (7) initializes the serial communication between the Arduino UNO board and the computer to display the serial monitor at 9,600 baud. The statements from (8) to (16) are used to display the humidity value in % and temperature value in °C in the serial monitor. The ambient humidity and temperature value greater than 90% and 25°C cause the condition of the statement (17) `if(h>90 && >25)` is true. The statements (18) and (19) will be executed and LED connected at Pin 8 of the Arduino board will turn on and "LED ON" will be displayed on the serial monitor. Otherwise, statement (20) will be executed to cause the execution of statements (21) and (22) and LED connected at Pin 8 of Arduino board will turn off and

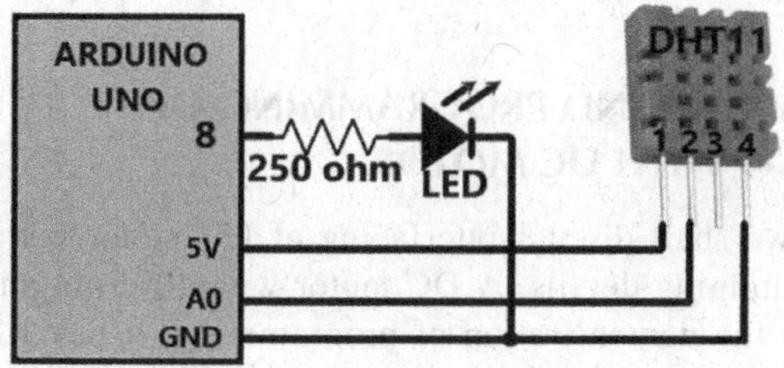

FIGURE 4.80 The interfacing of humidity and room temperature sensor DHT11 and LED with Arduino UNO board.

#include <dht.h>	*statement (1)*
dht DHT;	*statement (2)*
int analogInput=A0;	*statement (3)*
int LED=8;	*statement (4)*
void setup()	
{	
pinMode(analogInput,INPUT);	*statement (5)*
pinMode(LED,OUTPUT);	*statement (6)*
Serial.begin(9600);	*statement (7)*
}	
void loop()	
{	
DHT.read11(A0);	*statement (8)*
float t=DHT.temperature;	*statement (9)*
float h=DHT.humidity;	*statement (10)*
Serial.print("Humidity= ");	*statement (11)*
Serial.print(h);	*statement (12)*
Serial.println("%");	*statement (13)*
Serial.print("temperature= ");	*statement (14)*
Serial.print(t);	*statement (15)*
Serial.println("°C ");	*statement (16)*
if(h>90 && >25)	*statement (17)*
{	
digitalWrite(LED,HIGH);	*statement (18)*
Serial.println("LED ON");	*statement (19)*
}	
else	*statement (20)*
{	
digitalWrite(LED,LOW);	*statement (21)*
Serial.println("LED OFF");	*statement (22)*
}	
delay(2000);	*statement (23)*
}	

FIGURE 4.81 An Arduino UNO program to turn on the LED and display of ambient humidity and temperature value on the serial monitor if the humidity and temperature is greater than 90% and 25°C for the circuit diagram shown in Figure 4.80.

"LED OFF" will be displayed on the serial monitor. The statement (23) `delay(2000)` will generate a delay, and the whole process will be repeated after every 2 seconds.

4.11 INTERFACING AND PROGRAMMING OF ARDUINO WITH DC MOTOR

In this section, we shall discuss interfacing of DC motor with Arduino UNO board and programming details. A DC motor with 100 rpm and L293D motor driver is used in the demonstration of programming. It has 1.2 kg/cm torque, no-load current = 60 mA (Max), and load current = 300 mA (Max). The working principle of DC motor and L293D motor driver board is explained in Section 3.10 of Chapter 3.

COM6 (Arduino/Genuino Uno)

```
Humidity= 86.00%
temperature= 29.00°C
LED OFF
Humidity= 86.00%
temperature= 29.00°C
LED OFF
Humidity= 86.00%
temperature= 29.00°C
LED OFF
Humidity= 94.00%
temperature= 29.00°C
LED ON
Humidity= 90.00%
temperature= 29.00°C
LED OFF
```

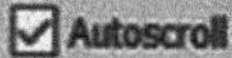

Newline

FIGURE 4.82 The screenshot of serial monitor of program shown in Figure 4.81 and interfacing circuit shown in Figure 4.80 to display the value of humidity and ambient temperature and status of LED using DHT11 on the serial monitor.

Program 4.40

Interface a 5 V DC motor with Arduino UNO board, and write a program to rotate it in the clockwise direction and display the information on serial monitor.

Solution

The interfacing of L293D-based motor driver board and 5 V DC motor with Arduino UNO board is shown in Figure 4.83a. The I/P 1 and I/P 2 pins of motor driver board are connected to Pins 9 and 8 of Arduino board. The O/P 1 and O/P 2 pins of motor driver board are connected to two terminals of 5 V DC motor. The 5 V and GND (ground) pins of motor driver board are connected to the 5 V and GND (ground) pins of Arduino board.

The interfacing schematic of L293D and 5 V DC motor with Arduino UNO board is shown in Figure 4.83b. The Pin 3 (Output 1) and Pin 6 (Output 2) of L293D motor driver IC are connected to the Terminal 1 and Terminal 2 of 5 V DC motor.

An Arduino UNO program to rotate the DC motor in the clockwise direction and display the information on the serial monitor for the circuit diagram shown in Figure 4.83a is shown in Figure 4.84.

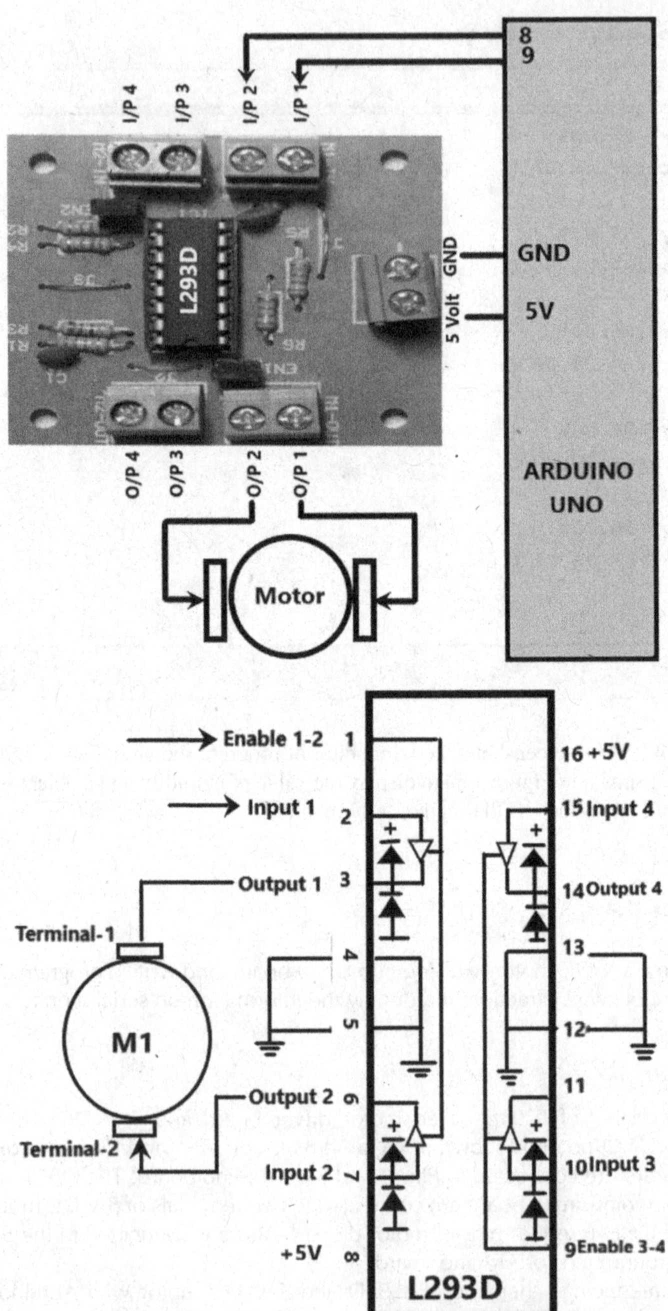

FIGURE 4.83 (a) The interfacing of a DC motor with Arduino UNO board using L293D. (b) The schematic of a DC motor with an Arduino UNO board using L293D

int Input1=9;	*statement (1)*
int Input2=8;	*statement (2)*
void setup()	
{	
pinMode(Input1,OUTPUT);	*statement (3)*
pinMode(Input2,OUTPUT);	*statement (4)*
Serial.begin(9600);	*statement (5)*
}	
void loop()	
{	
digitalWrite(Input1,HIGH);	*statement (6)*
digitalWrite(Input2,LOW);	*statement (7)*
Serial.println("Motor Rotating Clockwise");	*statement (8)*
}	

FIGURE 4.84 An Arduino UNO program to rotate the DC motor in the clockwise direction and display the information on the serial monitor for the circuit diagram shown in Figure 4.83 (a)

The screenshot of serial monitor of program shown in Figure 4.84 and interfacing circuit shown in Figure 4.83a to display the status of DC motor on the serial monitor is shown in Figure 4.85.

Description of the Program:

Using the statements (1) and (2), we give name "Input1" and "Input2" to Pins 9 and 8 of Arduino UNO, respectively. It is evident from Figure 4.81b that Pin 9 of Arduino is connected to Pin 2 (Input1) of L293D and Pin 8 of Arduino is connected to Pin 7 (Input2) of L293D. The signals will be sent out from Pins 9 and 8 of Arduino to the Input1 and Input2 pins of L293D; therefore, Arduino Pins 9 and 8 should be declared as output pins. The `pinMode(Input1,OUTPUT)` and `pinMode(Input2,OUTPUT)` functions in the statements(3) and (4) are used to declare Pins 9 and 8 of Arduino UNO board as an output pin. The statement (5) initializes the serial communication between the Arduino UNO board and the computer to display the serial monitor at 9,600 baud. The statement (6) `digitalWrite(-Input1,HIGH)` and the statement (7) `digitalWrite(Input2,LOW)` will send 5 and 0 V signals to Terminal 1 and Terminal 2 of DC motor. With the availability of 5 and 0 V signals on the Terminal 1 and Terminal 2, the DC motor starts rotating in the clockwise direction. The statement (8) `Serial.println("Motor Rotating Clockwise")` is used to display the message "Motor Rotating Clockwise" on the serial monitor.

Program 4.41

Interface a DC motor with Arduino UNO board, and write a program to rotate it in the clockwise direction for 5 seconds and stop it for 5 seconds and repeat it. The status of the motor should also be displayed on the serial monitor.

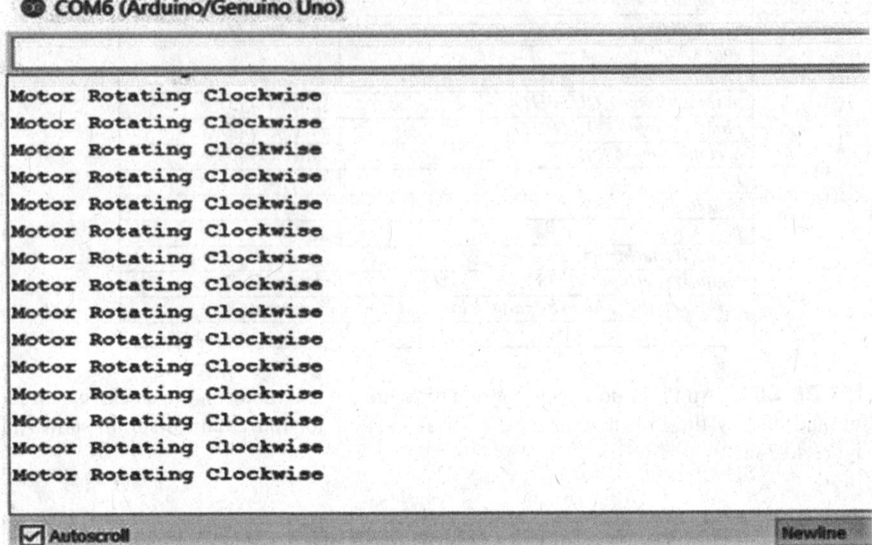

FIGURE 4.85 The screenshot of serial monitor of program shown in Figure 4.84 and interfacing circuit shown in Figure 4.83 (a) to display the status of DC motor on the serial monitor.

Solution

The interfacing and schematic diagram of a DC motor with an Arduino UNO board using L293D is shown in Figures 4.83(a) and 4.83 (b).

An Arduino UNO program to rotate the motor in the clockwise direction for 5 seconds and stop it for 5 seconds and repeat the whole sequence and at the same time display the status of the motor on the serial monitor for the circuit diagram shown in Figure 4.83(a) is shown in Figure 4.86.

The screenshot of serial monitor of program shown in Figure 4.84 and interfacing circuit shown in Figure 4.83(a) to display the status of DC motor on the serial monitor is shown in Figure 4.87.

Description of the Program:

The explanation of code up to statement (8) is the same as the explanation given for program as shown in Figure 4.84.

After the execution of statement (8), the motor rotates in the clockwise direction and the message "Motor Rotating Clockwise" will be displayed on the serial monitor.

The statement (9) `delay(5000)` will generate a delay of 5 seconds; thus, the motor rotates clockwise for 5 seconds.

The statement (10) `digitalWrite(Input1,LOW)` will send 0 V signal to the Terminal 1 of DC motor, which was having a signal of 5 V due to statement (6). Now, Terminal 1 of the motor will have 0 V, and Terminal 2 already has 0 V due to statement (7); therefore, the motor stops rotating.

int Input1=9;	statement (1)
int Input2=8;	statement (2)
void setup()	
{	
pinMode(Input1,OUTPUT);	statement (3)
pinMode(Input2,OUTPUT);	statement (4)
Serial.begin(9600);	statement (5)
}	
void loop()	
{	
digitalWrite(Input1,HIGH);	statement (6)
digitalWrite(Input2,LOW);	statement (7)
Serial.println("Motor Rotating Clockwise");	statement (8)
delay(5000);	statement (9)
digitalWrite(Input1,LOW);	statement (10)
Serial.println("Motor Stops");	statement (11)
delay(5000);	statement (12)
}	

FIGURE 4.86 An Arduino UNO program to rotate the motor in the clockwise direction for 5 seconds and stop it for 5 seconds and repeat the whole sequence and at the same time display the status of the motor on the serial monitor for the circuit diagram shown in Figure 4.83 (a)

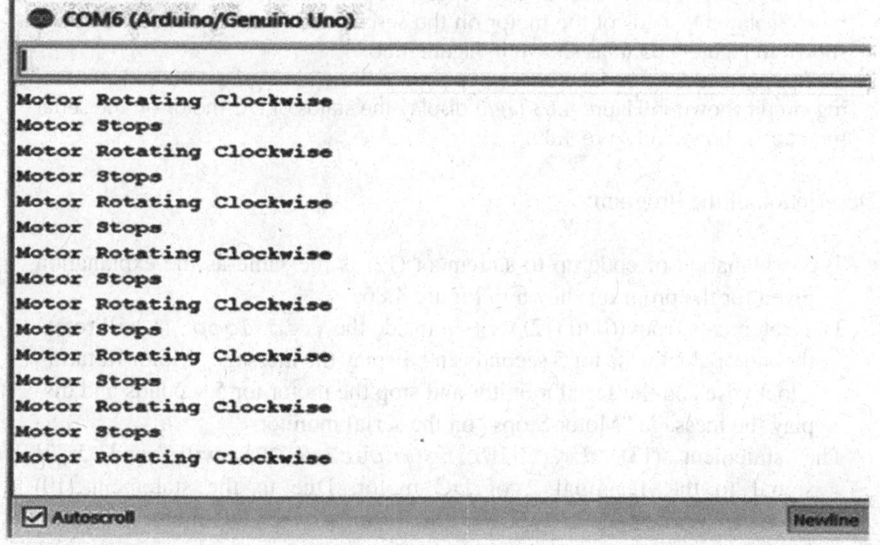

FIGURE 4.87 The screenshot of serial monitor of program shown in Figure 4.84 and interfacing circuit shown in Figure 4.83 (a) to display the status of DC motor on the serial monitor.

Now after the execution of statement (11) `Serial.println("Motor Stops")`, the message "Motor Stops" will be displayed on the serial monitor.

The statement (12) `delay(5000)` will generate a delay of 5 seconds; thus, the motor continues to stop for 5 seconds.

Since statement (12) is the last statement of the loop (), all the statements inside the loop, i.e., from (6) to (12), will be executed repeatedly.

Program 4.42

Interface a DC motor with Arduino UNO board, and write a program to rotate it in the clockwise direction for 5 seconds, stop it for 5 seconds, rotate it in an anticlockwise direction for 5 seconds, stop it for 5 seconds, and repeat the whole sequence. The status of the motor should also be displayed on the serial monitor.

Solution

The interfacing and schematic diagram of a DC motor with an Arduino UNO board using L293D is shown in Figure 4.83(a) and Figure 4.83 (b).

An Arduino UNO program to rotate the motor in the clockwise direction for 5 seconds, stop it for 5 seconds, rotate it in an anti-clockwise direction for 5 seconds, and stop it for 5 seconds and repeat the whole sequence and at the same time display the status of the motor on the serial monitor for the circuit diagram shown in Figure 4.83 (a) is shown in Figure 4.88.

The screenshot of serial monitor of program shown in Figure 4.88 and interfacing circuit shown in Figure 4.83 (a) to display the status of DC motor on the serial monitor is shown in Figure 4.89.

Description of the Program:

The explanation of code up to statement (12) is the same as the explanation given for the program shown in Figure 4.86.

The statements from (6) to (12) written inside the `void loop ()` will rotate the motor clockwise for 5 seconds and display the message "Motor Rotating Clockwise" on the serial monitor and stop the motor for 5 seconds and display the message "Motor Stops" on the serial monitor.

The statement (13) `digitalWrite(Input2,HIGH)` will send 5 V signal to the Terminal 2 of DC motor. Due to the statement (10) `digitalWrite(Input1,LOW)`, the Terminal 1 of DC motor is receiving 0 V; therefore, the motor starts rotating in an anti-clockwise direction.

The statement (14) `Serial.println("Motor Rotating Anti-Clockwise")` will display the message "Motor Rotating Anti-Clockwise" on the serial monitor. The statement (15) `delay(5000)` will generate a delay of 5 seconds; thus, the motor continues to rotate anticlockwise for 5 seconds. The statement (16) `digitalWrite(Input2,LOW)` will send 0 V signal to the Terminal 2 of DC motor. Due to the statement (10) `digitalWrite(Input1,LOW)` the Terminal 1 of DC motor is

int Input1=9;	*statement (1)*
int Input2=8;	*statement (2)*
void setup()	
{	
pinMode(Input1,OUTPUT);	*statement (3)*
pinMode(Input2,OUTPUT);	*statement (4)*
Serial.begin(9600);	*statement (5)*
}	
void loop()	
{	
digitalWrite(Input1,HIGH);	*statement (6)*
digitalWrite(Input2,LOW);	*statement (7)*
Serial.println("Motor Rotating Clockwise");	*statement (8)*
delay(5000);	*statement (9)*
digitalWrite(Input1,LOW);	*statement (10)*
Serial.println("Motor Stops");	*statement (11)*
delay(5000);	*statement (12)*
digitalWrite(Input2,HIGH);	*statement (13)*
Serial.println("Motor Rotating Anti-Clockwise");	*statement (14)*
delay(5000);	*statement (15)*
digitalWrite(Input2,LOW);	*statement (16)*
Serial.println("Motor Stops");	*statement (17)*
delay(5000);	*statement (18)*
}	

FIGURE 4.88 An Arduino UNO program to rotate the motor in the clockwise direction for 5 seconds, stop it for 5 seconds, rotate it in an anti-clockwise direction for 5 seconds and stop it for 5 seconds and repeat the whole sequence and at the same time display the status of the motor on the serial monitor for the circuit diagram shown in Figure 4.83 (a).

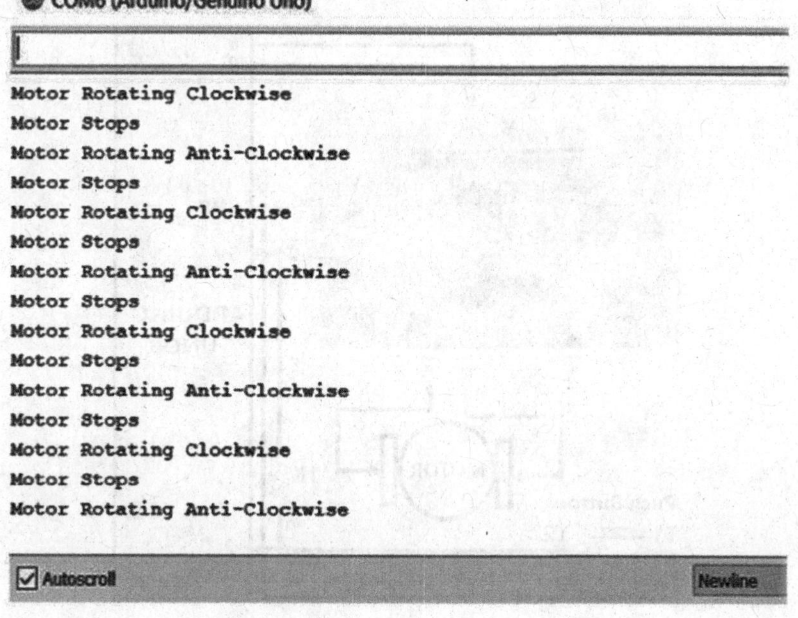

FIGURE 4.89 The screenshot of serial monitor of program shown in Figure 4.88 and interfacing circuit shown in Figure 4.83a to display the status of DC motor on the serial monitor.

receiving 0 V; therefore, the motor stops rotating. The statement (17) `Serial.println("Motor Stops")` will display the message "Motor Stops" on the serial monitor. The statement (18) `delay(5000)` will generate a delay of 5 seconds; thus, the motor continues to stop for 5 seconds. Since statement (18) is the last statement of the `void loop ()`, all the statements inside the loop, i.e., from (6) to (18), will be executed repeatedly.

Program 4.43

Interface a DC motor and a push-button switch with Arduino UNO board, and write a program to rotate the motor in the clockwise direction when the switch is not pressed and rotate the motor in an anti-clockwise direction when the switch is pressed. The status of the motor should also be displayed on the serial monitor.

Solution

The interfacing of L293D-based motor driver board, 5 V DC motor, and push-button switch with Arduino UNO board is shown in Figure 4.90. The I/P 1 and I/P 2 pins of motor driver board are connected to the Pins 9 and 8 of Arduino board. The O/P 1 and O/P 2 pins of motor driver board are connected to two terminals of 5 V DC motor. The 5 V and GND (ground) pins of motor driver board are connected to the 5 V and GND (ground) pins of Arduino board. The Terminal T1 of the push button is connected to the GND (ground) pin of Arduino board, and the Terminal T2 is connected to the one terminal of 1 KΩ resistor. The other terminal of 1 KΩ

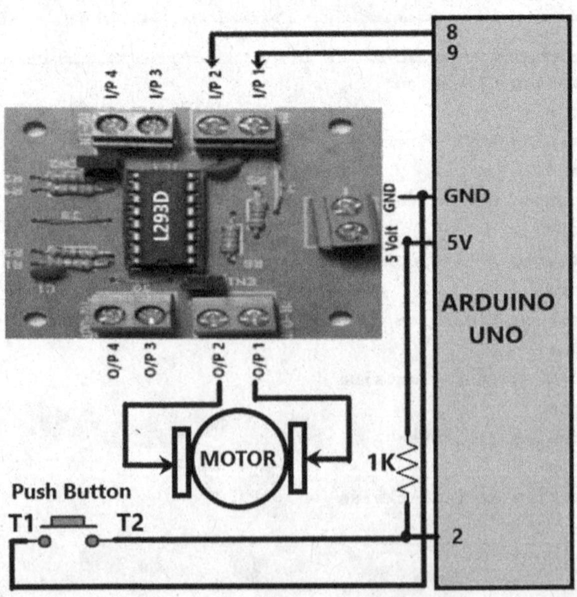

FIGURE 4.90 The interfacing of a DC motor and a push-button switch with Arduino UNO board.

resistor is connected to the 5 V pin of Arduino board. The junction of Terminal T2 of switch and 1 KΩ resistor is extended and connected to Pin 2 of Arduino board.

An Arduino UNO program to rotate the motor in the clockwise direction when the switch is not pressed and rotate the motor in an anti-clockwise direction when the switch is pressed and at the same time display the status of the motor on the serial monitor for the circuit diagram shown in Figure 4.90 is shown in Figure 4.91.

The screenshot of serial monitor of program shown in Figure 4.91 and interfacing circuit shown in Figure 4.90 to display the status of DC motor on the serial monitor is shown in Figure 4.92.

Description of the Program:

By using the statements (1), (2), and (3), we give the name "Input1", "Input2", and "pushButton" to Pins 9, 8, and 2 of Arduino UNO, respectively. It is evident from Figure 4.90 that Pins 9 and 8 of Arduino are connected to Pin 2 (Input1) and Pin 7 (Input2) of L293D. The signals will be sent out from Pins 9 and 8 of Arduino to Pins 2 and 7 of L293D; therefore, Arduino Pins 9 and 8 should be declared as output pins. The `pinMode(Input1,OUTPUT)`

int Input1=9;	*statement (1)*
int Input2=8;	*statement (2)*
int pushButton=2;	*statement (3)*
void setup()	
{	
pinMode(Input1,OUTPUT);	*statement (4)*
pinMode(Input2,OUTPUT);	*statement (5)*
pinMode(pushButton,INPUT);	*statement (6)*
Serial.begin(9600);	*statement (7)*
}	
void loop()	
{	
int buttonState=digitalRead(pushButton);	*statement (8)*
if (buttonState==1)	*statement (9)*
{	
digitalWrite(Input1,HIGH);	*statement (10)*
digitalWrite(Input2,LOW);	*statement (11)*
Serial.println("Motor Rotating Clockwise");	*statement (12)*
delay(100);	*statement (13)*
}	
else if (buttonState==0)	*statement (14)*
{	
digitalWrite(Input1, LOW);	*statement (15)*
digitalWrite(Input2, HIGH);	*statement (16)*
Serial.println ("Motor Rotating Anti-Clockwise ");	*statement (17)*
delay(100);	*statement (18)*
}	
}	

FIGURE 4.91 An Arduino UNO program to rotate the motor in the clockwise direction when the switch is not pressed and rotate the motor in an anti-clockwise direction when the switch is pressed and at the same time display the status of the motor on the serial monitor for the circuit diagram shown in Figure 4.90.

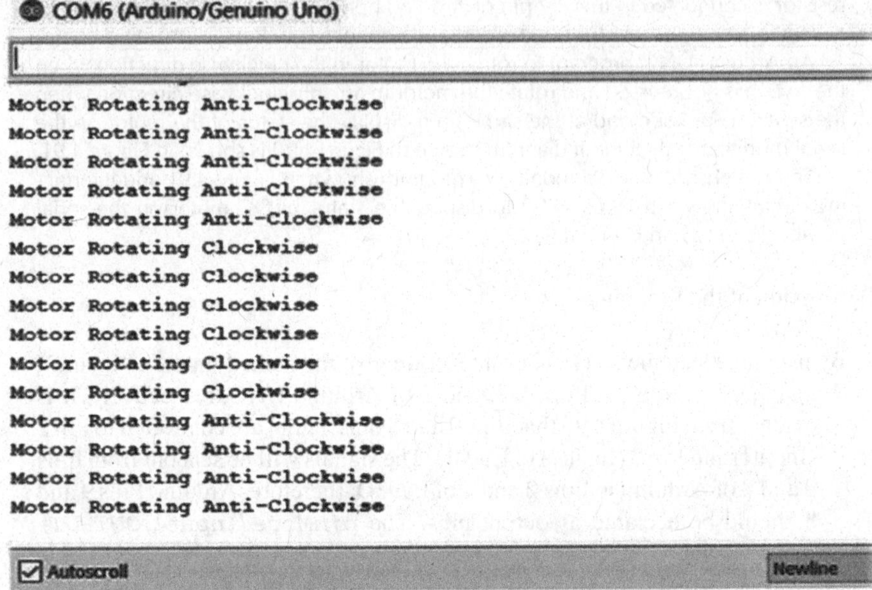

FIGURE 4.92 The screenshot of serial monitor of program shown in Figure 4.91 and interfacing circuit shown in Figure 4.90 to display the status of DC motor on the serial monitor.

and *pinMode(Input2,OUTPUT)* functions in the statements (4) and (5) are used to declare Pins 9 and 8 of Arduino UNO board as an output pin. The *pinMode(pushButton, INPUT)* function in the statement (6) is used to declare Pin 2 of the Arduino UNO board as an input pin. The statement (7) initializes the serial communication between the Arduino UNO board and the computer to display the serial monitor at 9,600 baud. By using the statement (8) *int buttonState=digitalRead(pushButton)*, the digital value of "pushButton" (Pin 2) will be read and assigned to the variable "buttonState". As per Figure 4.90 if the push button is not pressed, then Logic 1 (5 V) will be assigned to variable "buttonState", and if the push button is pressed, then Logic 0 (0 V) will be assigned to variable "buttonState". If the push-button switch is not pressed, the condition of the statement (9) *if (buttonState==1)* is true and it will cause statements (10) to (13) to be executed, and the motor rotates clockwise; otherwise, statement (14) *else if (buttonState==0)* is true. It will cause statements (15) to (18) to be executed, and the motor rotates anti-clockwise.

Program 4.44

Interface a DC motor and a push-button switch with Arduino UNO board, and write a program to rotate the motor in two different speeds in the clockwise direction. When the switch is not pressed, the motor rotates in highest speed, and the motor rotates in medium speed when the switch is pressed. The status of the motor should also be displayed on the serial monitor.

Solution

The interfacing of L293D-based motor driver board, 5 V DC motor, and push-button switch with Arduino UNO board is shown in Figure 4.93. The I/P 1 and I/P 2 pins of motor driver board are connected to Pins 10 and 9 of Arduino board. The O/P 1 and O/P 2 pins of motor driver board are connected to two terminals of 5 V DC motor. The 5 V and GND (ground) pins of motor driver board are connected to the 5 V and GND (ground) pins of Arduino board. The Terminal T1 of the push button is connected to the GND (ground) pin of Arduino board, and Terminal T2 is connected to the one terminal of 1 KΩ resistor. The other terminal of 1 KΩ resistor is connected to the 5 V pin of Arduino board. The junction of Terminal T2 of switch and 1 KΩ resistor is extended and connected to Pin 2 of Arduino board.

The interfacing schematic of L293D and 5 V DC motor with Arduino UNO board is shown in Figure 4.94. The Pin 3 (Output 1) and Pin 6 (Output 2) of L293D motor driver IC are connected to Terminal 1 and Terminal 2 of 5 V DC motor, respectively.

In this example, we have to rotate the motor in two different speeds; therefore, we require to interface motor with PWM-capable pins. We have interfaced the motor using the L293D motor driver board with Pins 10 and 9 of Arduino UNO board, PWM pins. These pins are connected to the Input1 and Input2 pins of L293D IC.

An Arduino UNO program is used to rotate the motor in two different speeds in the clockwise direction. A program is shown in Figure 4.95 which rotates the

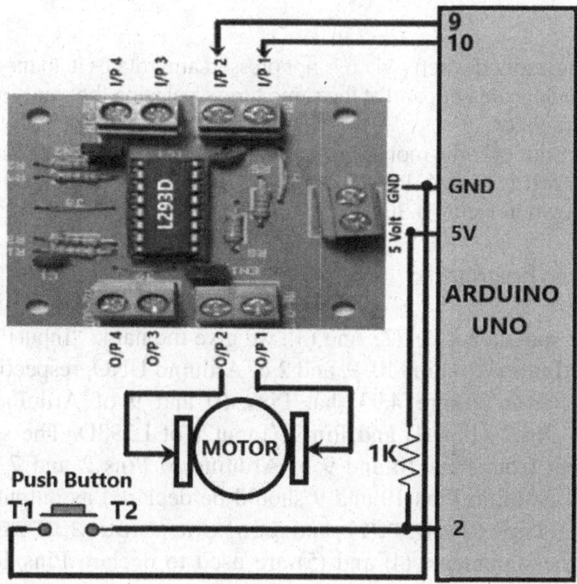

FIGURE 4.93 The interfacing of a DC motor, push-button switch with Arduino UNO board using L293D for generating PWM effect.

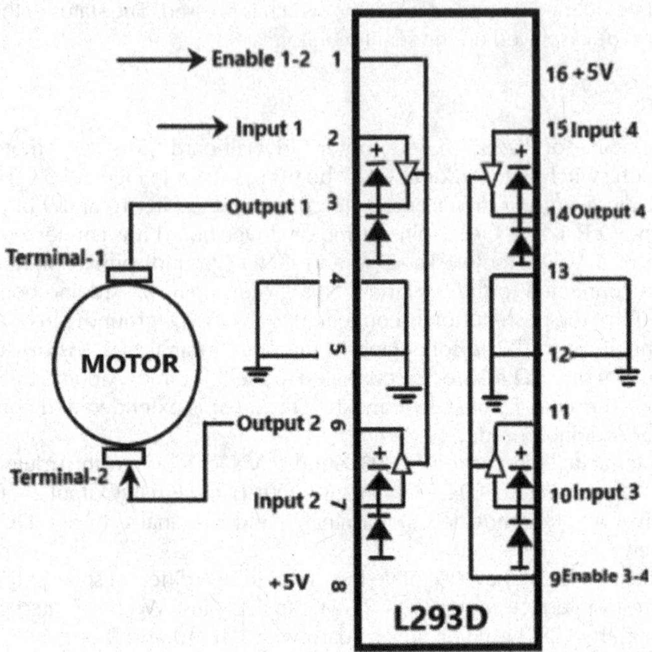

FIGURE 4.94 The interfacing schematic of L293D and 5 V DC motor with Arduino UNO board is shown in figure. The Pin 3 (Output 1) and Pin 6 (Output 2) of L293D motor driver IC are connected to Terminal 1 and Terminal 2 of 5 V DC motor.

motor in highest speed when switch is not pressed and rotates it in medium speed when the switch is pressed, and at the same time displaying the status of the motor on the serial monitor.

The screenshot of serial monitor of program shown in Figure 4.95 and interfacing circuit shown in Figure 4.93 to display the status of DC motor on the serial monitor is shown in Figure 4.96.

Description of the Program:

By using the statements (1), (2), and (3), we give the name "Input1", "Input2", and "pushButton" to Pins 10, 9, and 2 of Arduino UNO, respectively.

It is evident from Figure 4.93 that Pins 10 and 9 of Arduino are connected to Pin 2 (Input1) and Pin 7 (Input2) of L293D. The signals will be sent out from Pins 10 and 9 of Arduino to Pins 2 and 7 of L293D; therefore, Arduino Pins 10 and 9 should be declared as output pins. The *pinMode(Input1,OUTPUT)* and *pinMode(Input2,OUTPUT)* functions in the statements (4) and (5) are used to declare Pins 10 and 9 of Arduino UNO board as an output pin. The *pinMode(pushButton, INPUT)* function in the statement (6) is used to declare Pin 2 of the Arduino UNO board as an input pin.

int Input1=10;	*statement (1)*
int Input2=9;	*statement (2)*
int pushButton=2;	*statement (3)*
void setup()	
{	
pinMode(Input1,OUTPUT);	*statement (4)*
pinMode(Input2,OUTPUT);	*statement (5)*
pinMode(pushButton,INPUT);	*statement (6)*
Serial.begin(9600);	*statement (7)*
}	
void loop()	
{	
int buttonState=digitalRead(pushButton);	*statement (8)*
if (buttonState==1)	*statement (9)*
{	
analogWrite(Input1,255);	*statement (10)*
analogWrite (Input2,0);	*statement (11)*
Serial.println("Motor Rotating at Highest Speed");	*statement (12)*
delay(100);	*statement (13)*
}	
else if (buttonState==0)	*statement (14)*
{	
analogWrite (Input1,127);	*statement (15)*
analogWrite (Input2, 0);	*statement (16)*
Serial.println ("Motor Rotating at Medium Speed ");	*statement (17)*
delay(100);	*statement (18)*
}	
}	

FIGURE 4.95 An Arduino UNO program to rotate the motor in two different speeds in the clockwise direction. When the switch is not pressed, the motor rotates in highest speed, and it rotates in medium speed when the switch is pressed and at the same time display the status of the motor on the serial monitor for the circuit diagram shown in Figure 4.93.

The statement (7) initializes the serial communication between the Arduino UNO board and the computer to display the serial monitor at 9,600 baud.

By using the statement (8) `int buttonState=digitalRead(pushButton)`, the digital value of "pushButton" (pin number 2) will be read and assigned to the variable "buttonState". As per Figure 4.93 if the push button is not pressed, then Logic 1 (5 V) will be assigned to variable "buttonState", and if push button is pressed, then Logic 0 (0 V) will be assigned to variable "buttonState".

Suppose the push-button switch has not pressed, the condition of the statement (9) `if (buttonState==1)` is true and it will cause statements (10) to (13) to be executed. The statement (10) `analogWrite(Input1,255)` will write the value 255 to Pin 10, and by doing so, Pin 10 will generate 5 V. The statement (11) `analogWrite(Input2, 0)` will write the value 0 to Pin 9, and by doing so, Pin 9 will generate 0 V.

The 5 V and 0 V outputs of Pins 10 and 9 are connected to Input1 and Input2 pins of L293D board. The Output1 and Output2 pins of L293D board will

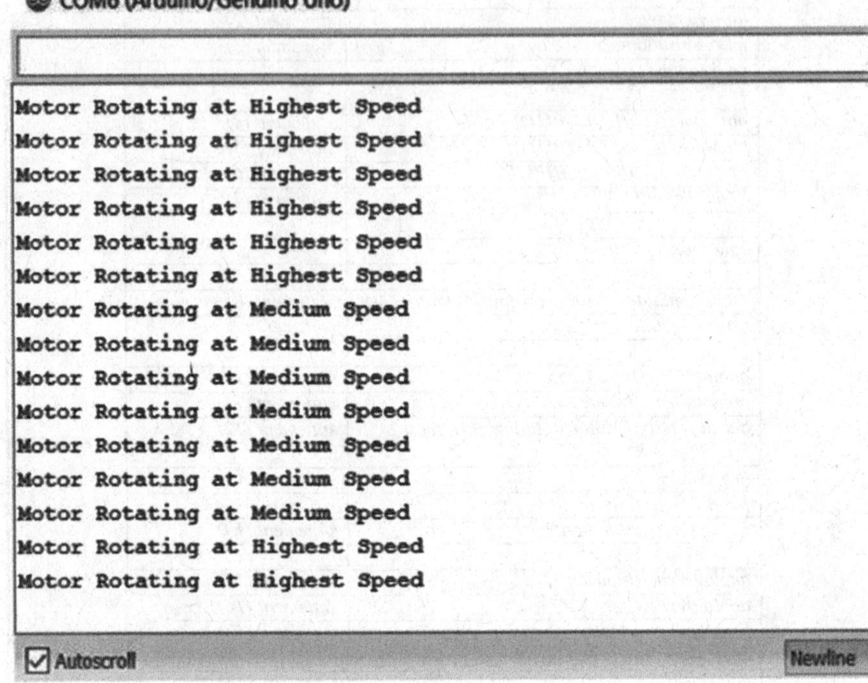

FIGURE 4.96 The screenshot of serial monitor of program shown in Figure 4.95 and interfacing circuit shown in Figure 4.93 to display the status of DC motor on the serial monitor.

send out 5 V and 0 V and connected to the Terminal 1 and Terminal 2 of the motor (Refer Figure 4.94). Since the Terminal 1 and Terminal 2 of the motor are connected to 5 V and 0 V, motor run with the fastest speed and statement (12) will display "Motor Rotating at Highest Speed" on the serial monitor.

Suppose the push-button switch has not pressed, the condition of the statement (14) `if (buttonState==0)` is true and it will cause statements (15) to (18) to be executed. The statement (15) `analogWrite(Input1,127)` will write the value 127 to Pin 10, and by doing so, Pin 10 will generate 2.5 V. The statement (16) `analogWrite(Input2,0)` will write the value 0 to Pin 9, and by doing so, Pin 9 will generate 0 V.

The 2.5 V and 0 V outputs of Pins 10 and 9 are connected to Input1 and Input2 pins of L293D board. The Output1 and Output2 pins of L293D board will send out 2.5 and 0 V, and are connected to Terminal 1 and Terminal 2 of the motor (Refer Figure 4.94). Since the Terminal 1 and Terminal 2 of the motor are connected to 2.5 V and 0 V, motor runs with a medium speed and statement (17) will display "Motor Rotating at Medium Speed" on the serial monitor.

4.12 INTERFACING AND PROGRAMMING OF ARDUINO WITH HIGH-VOLTAGE DEVICE AND RELAY

This section shall discuss interfacing of high-voltage devices with Arduino UNO board using relay and programming details. Relay is an electromechanical device controlled by small voltage/current and switch on or off high-voltage/current device. Using relay, we can control the AC-operated devices by using a small DC signal. The working principle of relay is explained in Section 3.11 of Chapter 3.

Program 4.45

Interface bulb and relay with Arduino UNO board, and write a program to turn on and off the bulb for 5 and 8 seconds, respectively, and display the serial monitor's information.

Solution

The interfacing of 5 V relay board and 220 V AC-operated bulb with Arduino UNO board is shown in Figure 4.97. The 220 V AC-operated bulb is connected to NC (normally close) and COM (also called as pole) pins of relay. The 5 V and GND (ground) pins of relay are connected to 5 V and GND (ground) pins of Arduino board. The IN pin of relay is connected to Pin 8 of Arduino board.

The schematic of a 220 V AC-operated bulb connected to COM and NC (normally close) terminals of relay is shown in Figure 4.98. The bulb is on when the relay is not triggered by applying 0 V at Pin A of relay connected to Pin 8 of Arduino board.

The schematic of a 220 V AC-operated bulb connected to COM and NC (normally close) terminals of relay is shown in Figure 4.99. The bulb is off when the relay is triggered by applying 5 V at Pin A of relay connected to Pin 8 of Arduino board.

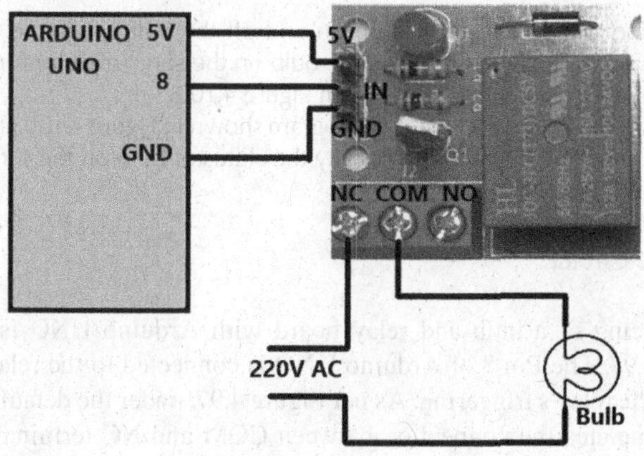

FIGURE 4.97 The interfacing of 5 V relay board and 220 V AC-operated bulb with Arduino UNO board.

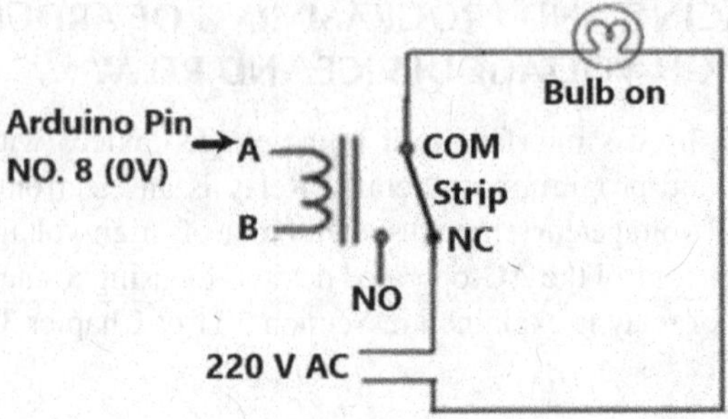

FIGURE 4.98 The schematic of a 220 V AC-operated bulb connected to COM and NC (normally close) terminals of relay. The bulb is on by not triggering the relay.

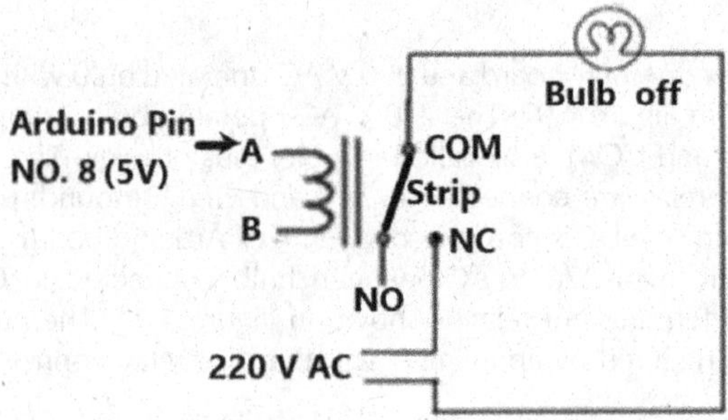

FIGURE 4.99 The schematic of a 220 V AC-operated bulb connected to COM and NC (normally close) terminals of relay. The bulb is off by triggering the relay.

An Arduino UNO program to turn on and off the bulb for 5 and 8 seconds, respectively, and display the status of the bulb on the serial monitor for the circuit diagram shown in Figure 4.97 is shown in Figure 4.100.

The screenshot of serial monitor of program shown in Figure 4.100 and interfacing circuit shown in Figure 4.97 to display the status of bulb on the serial monitor is shown in Figure 4.101.

Working of the Circuit:

The interfacing of a bulb and relay board with Arduino UNO is shown in Figure 4.97. The Pin 8 of Arduino UNO is connected to the relay board to control the relay's triggering. As per Figure 4.97, under the default condition strip completes the connection between COM and NC terminals of relay, and this will cause the bulb to turn on (see Figure 4.98). If Pin 8 generates 5 V, then the relay will be triggered, and the strip will complete the connection between COM and NO terminals of relay and this will cause the bulb to turn off (see Figure 4.99).

int controlBulb=8;	statement (1)
void setup()	
{	
pinMode(controlBulb,OUTPUT);	statement (2)
Serial.begin(9600);	statement (3)
}	
void loop()	
{	
digitalWrite(controlBulb,LOW);	statement (4)
Serial.println("Bulb ON");	statement (5)
delay(5000);	statement (6)
digitalWrite(controlBulb,HIGH);	statement (7)
Serial.println("Bulb OFF");	statement (8)
delay(8000);	statement (9)
}	

FIGURE 4.100 An Arduino UNO program to turn on and off the bulb for 5 and 8 seconds, respectively, and display the status of the bulb on the serial monitor for the circuit diagram shown in Figure 4.97.

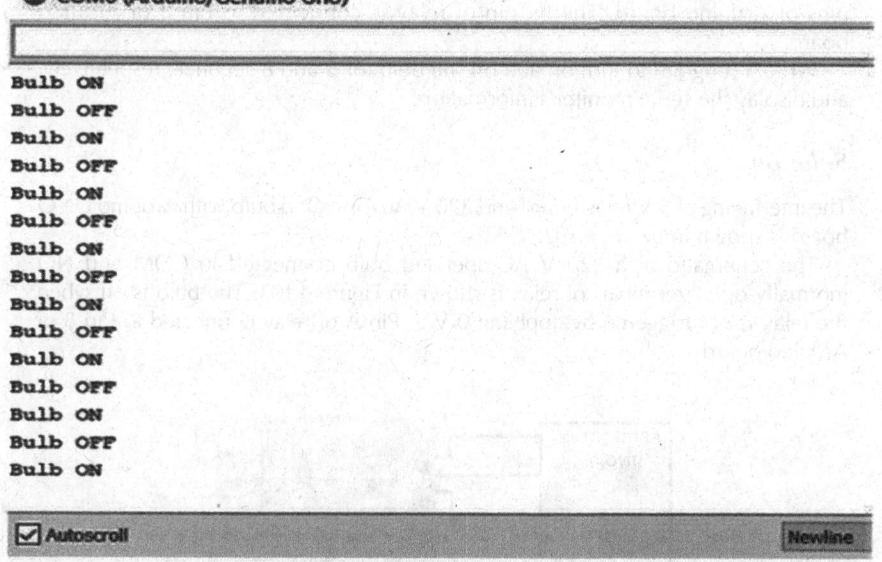

FIGURE 4.101 The screenshot of serial monitor of program shown in Figure 4.100 and interfacing circuit shown in Figure 4.97 to display the status of bulb on the serial monitor.

Description of the Program:

Using the statements (1) and (2), Pin 8 is declared as an output pin. The statement (3) initializes the serial communication between the Arduino UNO board and the computer to display the serial monitor at 9,600 baud.

The statement (4) `digitalWrite(controlBulb, LOW)` will output 0 V at Pin 8 of Arduino under this condition the relay will not trigger, and the bulb will remain on (Figure 4.98). The statement (5) will display "Bulb ON", and the statement (6) will generate a delay of 5 seconds.

The statement (7) `digitalWrite(controlBulb, HIGH)` will output 5 V at Pin 8 of Arduino under this condition, the relay will trigger, and the bulb will turn off (Figure 4.99). The statement (8) will display "Bulb OFF", and the statement (9) will generate a delay of 8 seconds.

Program 4.46

The interfacing of 5 V relay board and 220 V AC-operated bulb with Arduino UNO board is shown in Figure 4.102. The 220 V AC-operated bulb is connected to NO (normally open) and COM (also called as pole) pins of relay. The 5 V and GND (ground) pins of relay are connected to 5 V and GND (ground) pins of Arduino board. The IN pin of relay is connected to Pin 8 of Arduino board.

Write a program to turn on and off the bulb for 5 and 8 seconds, respectively, and display the serial monitor's information.

Solution

The interfacing of 5 V relay board and 220 V AC-operated bulb with Arduino UNO board is shown in Figure 4.102.

The schematic of a 220 V AC-operated bulb connected to COM and NO (normally open) terminals of relay is shown in Figure 4.103. The bulb is off when the relay is not triggered by applying 0 V at Pin A of relay connected to Pin 8 of Arduino board.

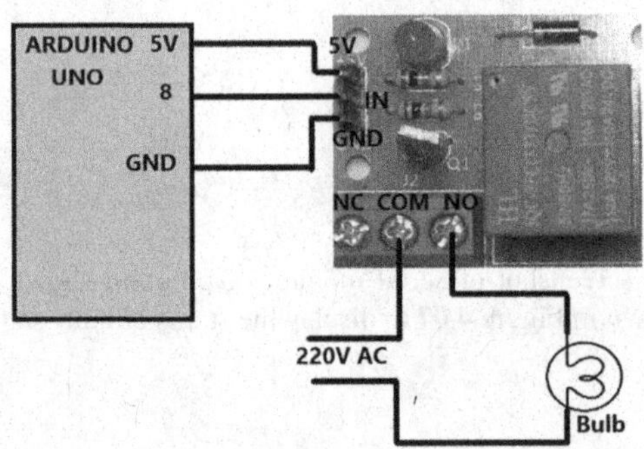

FIGURE 4.102 The interfacing of a bulb and relay board with Arduino UNO for Program 4.46.

The schematic of a 220 V AC-operated bulb connected to COM and NO (normally open) terminals of relay is shown in Figure 4.104. The bulb is on when the relay is triggered by applying 5 V at Pin A of relay connected to Pin 8 of Arduino board.

An Arduino UNO program to turn on and off the bulb for 5 and 8 seconds, respectively, and display the status of the bulb on the serial monitor for the circuit diagram shown in Figure 4.102 is shown in Figure 4.105.

The screenshot of serial monitor of program shown in Figure 4.105 and interfacing circuit shown in Figure 4.102 to display the status of bulb on the serial monitor is shown in Figure 4.106.

Working of the Circuit:

The interfacing of a bulb and relay board with Arduino UNO is shown in Figure 4.102. The Pin 8 of Arduino UNO is connected to the relay board to control the relay's triggering. As per Figure 4.102, under the default condition, strip is connected between COM and NC terminals of relay and the bulb will be turned off (see Figure 4.103). If Pin 8 generates 5 V, then the relay will be triggered, and the strip will complete the connection between COM and NO terminals of relay and the bulb will be turned on (see Figure 4.104).

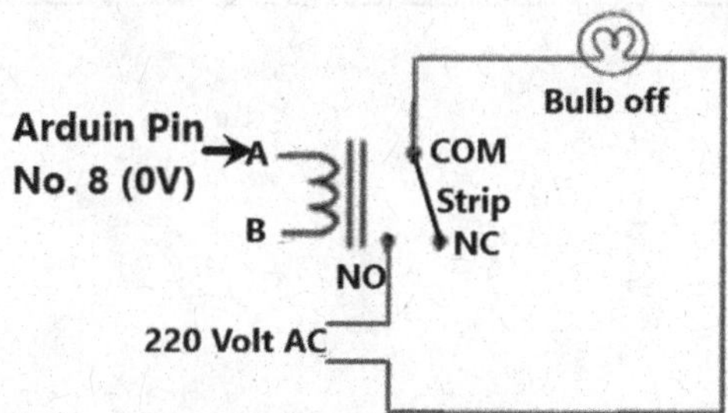

FIGURE 4.103 The schematic of a 220 V AC-operated bulb connected to COM and NO (normally open) terminals of relay. The bulb is off by not triggering the relay.

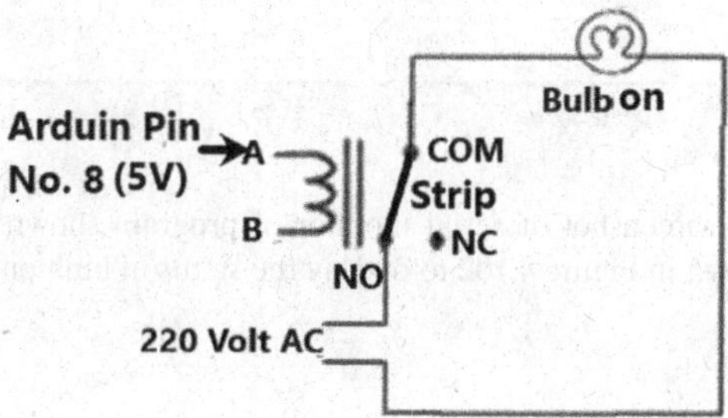

FIGURE 4.104 The schematic of a 220 V AC-operated bulb connected to COM and NO (normally open) terminals of relay. The bulb is on by triggering the relay.

int controlBulb=8;	*statement (1)*
void setup()	
{	
pinMode(controlBulb,OUTPUT);	*statement (2)*
Serial.begin(9600);	*statement (3)*
}	
void loop()	
{	
digitalWrite(controlBulb,HIGH);	*statement (4)*
Serial.println("Bulb ON");	*statement (5)*
delay(5000);	*statement (6)*
digitalWrite(controlBulb,LOW);	*statement (7)*
Serial.println("Bulb OFF");	*statement (8)*
delay(8000);	*statement (9)*
}	

FIGURE 4.105 An Arduino UNO program to turn on and off the bulb for 5 and 8 seconds, respectively, and display the status of the bulb on the serial monitor for the circuit diagram shown in Figure 4.102.

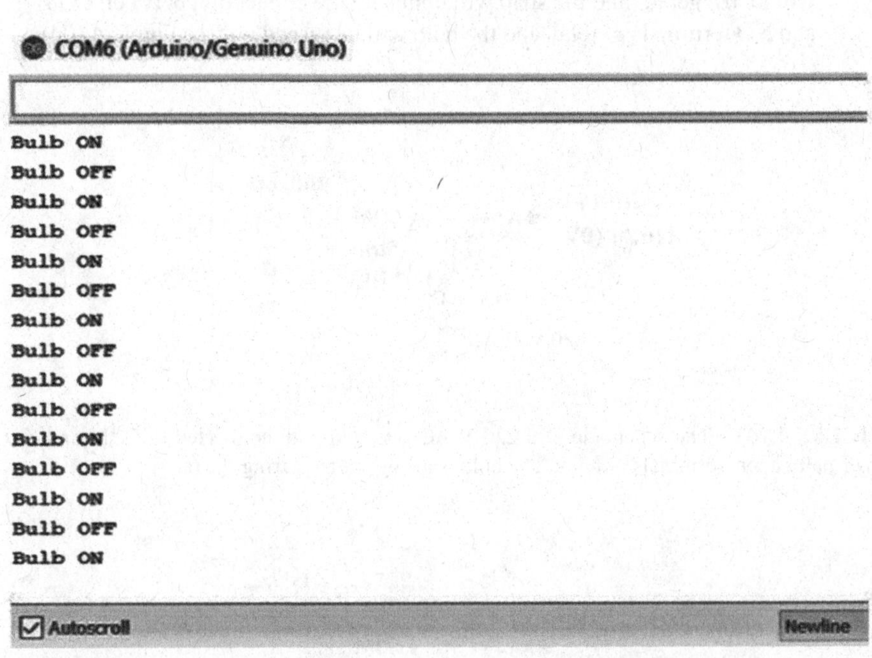

FIGURE 4.106 The screenshot of serial monitor of program shown in Figure 4.105 and interfacing circuit shown in Figure 4.102 to display the status of bulb on the serial monitor.

Description of the Program:

The program's description is self-explanatory and similar to the description of the program shown in Figure 4.100 except statements (4) and (7). Due to the change in the connection between bulb and pins of the relay, to turn on the bulb, we have to trigger the relay; therefore, statement (4) of Program shown in Figure 4.100 will be replaced by `digitalWrite(controlBulb, HIGH)`. To turn off the bulb, we have to turn off the relay; therefore, statement (7) of program shown in Figure 4.100 will be replaced by `digitalWrite(controlBulb, LOW)`.

4.13 INTERFACING AND PROGRAMMING OF ARDUINO UNO WITH LIGHT-DEPENDENT RESISTOR (LDR)

In this section, we shall discuss the interfacing of light-dependent resistor (LDR) with Arduino UNO board and programming details. The LDR is a light-sensitive device whose resistance varies with a change in the intensity of light falling on it. The working principle of LDR sensor is explained in Section 3.12 of Chapter 3.

Program 4.47

Interface an LDR, bulb, and relay with Arduino UNO board, and write a program to turn on the bulb when night and turn off the bulb when day, and display the information on serial monitor.

Solution

The interfacing of 5 V relay board, 220 V AC-operated bulb, and LDR with Arduino UNO board is shown in Figure 4.107. The 220 V AC-operated bulb is connected to NC (normally close) and COM (also called as pole) pins of relay. The 5 V and GND (ground) pins of relay are connected to 5 V and GND (ground) pins of Arduino board. The IN pin of relay is connected to Pin 8 of Arduino board. The Terminal T1 of the LDR is connected to the 5 V pin of Arduino board, and the Terminal T2 is connected to the one terminal of 10 KΩ resistor. The other terminal of 10 KΩ resistor is connected to the GND (ground) pin of Arduino board. The junction of Terminal T2 of LDR and 10 KΩ resistor is extended and connected to the A0 pin of Arduino board.

The schematic of a 220 V AC-operated bulb connected to COM and NC (normally close) terminals of relay is shown in Figure 4.108. The relay is not triggered in night by applying 0 V at Pin A of relay connected to Pin 8 of Arduino board and bulb is on.

The schematic of a 220 V AC-operated bulb connected to COM and NC (normally close) terminals of relay is shown in Figure 4.109. The relay is triggered in day by applying 5 V at Pin A of relay connected to Pin 8 of Arduino board and the bulb is off.

An Arduino UNO program to turn on the bulb in the night and turn off in the day and display the status of the bulb on the serial monitor for the circuit diagram shown in Figure 4.107 is shown in Figure 4.111.

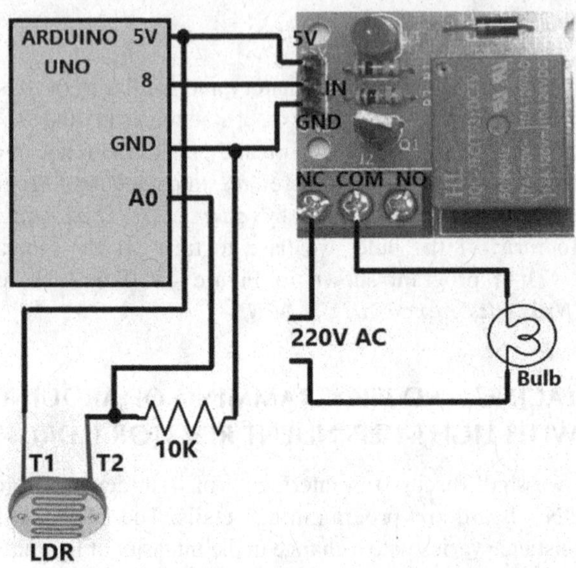

FIGURE 4.107 The interfacing of an LDR, bulb, and relay with Arduino UNO board.

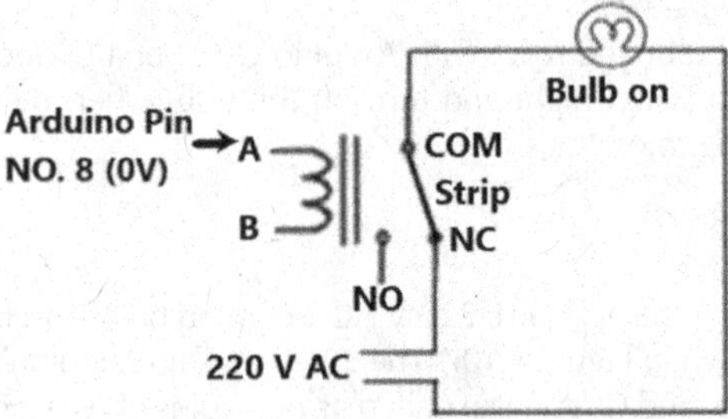

FIGURE 4.108 The schematic of a 220 V AC-operated bulb connected to COM and NC (normally close) terminals of relay. The bulb is on in night by not triggering the relay.

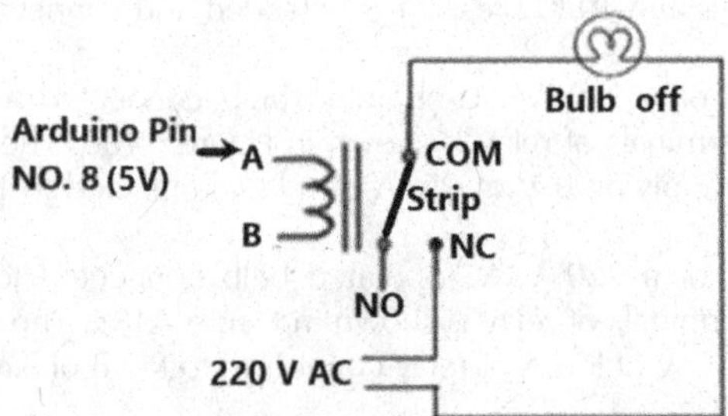

FIGURE 4.109 The schematic of a 220 V AC-operated bulb connected to COM and NC (normally close) terminals of relay. The bulb is off in day by triggering the relay.

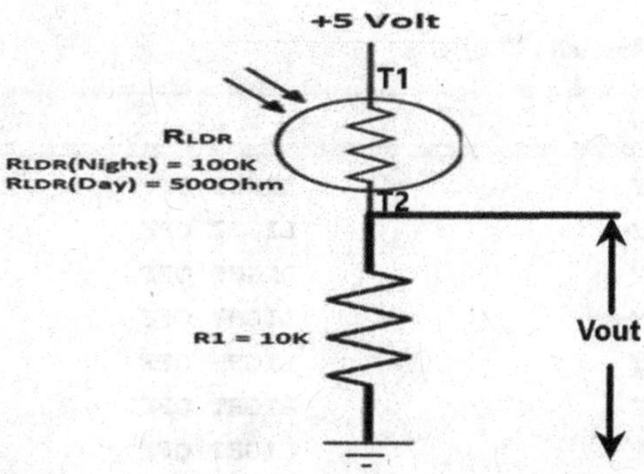

FIGURE 4.110 The LDR-based control circuit to control the on and off of bulb depending upon the brightness.

int bulb=8;	*statement (1)*
int analogInput=A0;	*statement (2)*
void setup()	
{	
pinMode(bulb,OUTPUT);	*statement (3)*
pinMode(analogInput,INPUT);	*statement (4)*
Serial.begin(9600);	*statement (5)*
}	
void loop()	
{	
int step=analogRead(analogInput);	*statement (6)*
if (step<=300)	*statement (7)*
{	
digitalWrite(bulb,LOW);	*statement (8)*
Serial.print(step);	*statement (9)*
Serial.print(" NIGHT");	*statement (10)*
Serial.println(" LIGHT ON");	*statement (11)*
delay(2000);	*statement (12)*
}	
else	*statement (13)*
{	
digitalWrite(bulb,HIGH);	*statement (14)*
Serial.print(step);	*statement (15)*
Serial.print(" DAY");	*statement (16)*
Serial.println(" LIGHT OFF");	*statement (17)*
delay(2000);	*statement (18)*
}	
}	

FIGURE 4.111 An Arduino UNO program to turn on the bulb in the night and turn off in the day and display the status of the bulb on the serial monitor for the circuit diagram shown in Figure 4.107.

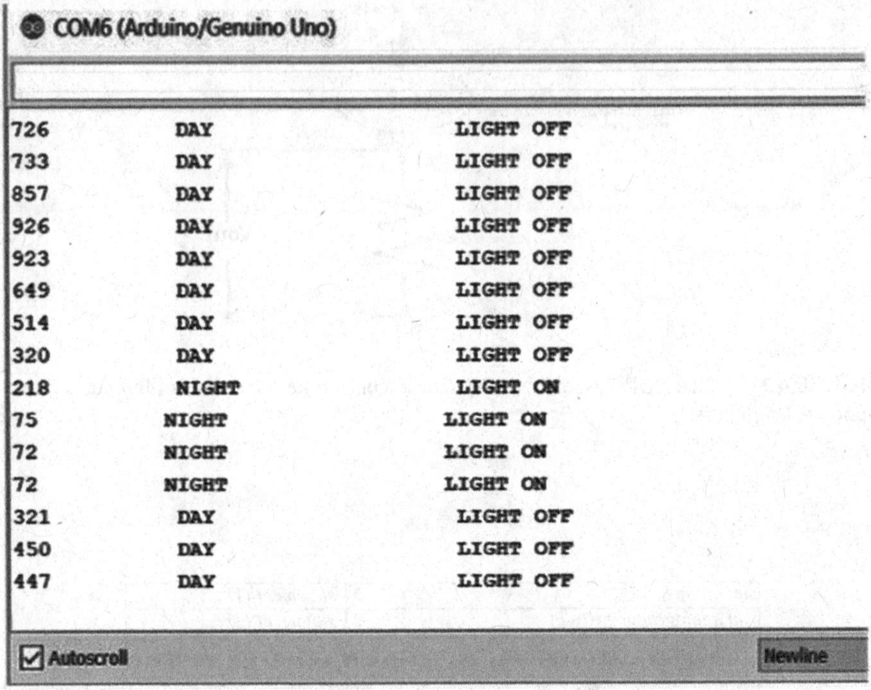

```
COM6 (Arduino/Genuino Uno)

726     DAY      LIGHT OFF
733     DAY      LIGHT OFF
857     DAY      LIGHT OFF
926     DAY      LIGHT OFF
923     DAY      LIGHT OFF
649     DAY      LIGHT OFF
514     DAY      LIGHT OFF
320     DAY      LIGHT OFF
218     NIGHT    LIGHT ON
75      NIGHT    LIGHT ON
72      NIGHT    LIGHT ON
72      NIGHT    LIGHT ON
321     DAY      LIGHT OFF
450     DAY      LIGHT OFF
447     DAY      LIGHT OFF

Autoscroll                                    Newline
```

FIGURE 4.112 The screenshot of serial monitor of program shown in Figure 4.111 and interfacing circuit shown in Figure 4.107 to display the status of bulb on the serial monitor.

The screenshot of serial monitor of program shown in Figure 4.111 is shown in Figure 4.112.

Working of the Circuit:

The interfacing of an LDR, bulb, and relay with Arduino UNO board is shown in Figure 4.107. The Pin 8 of Arduino UNO is connected to the relay board to control the relay's triggering. The bulb has to be turned on and off depending upon the brightness. If it is the night, the bulb has to be turned on, and if it is the day, the bulb has to be turned off. Turning on and off of bulb is done by using the relay. When the relay does not trigger, the strip completes the connection between COM and NC terminals of relay, and this will cause the bulb to be on (see Figure 4.108). If the relay is triggered, then strip will complete the connection between COM and NO terminals of relay, and this will cause the bulb to be off (see Figure 4.109).

The LDR-based control circuit to control the turning of on and off of the bulb depending upon the brightness is shown in Figure 4.110. The Terminal T1 of the LDR is connected to +5 V, and T2 is connected to one end of 10 KΩ resistor. The other end of the 10 KΩ resistor is connected to GND (ground). The output voltage (Vout) is taken across 10 KΩ resistor.

The LDR is chosen such that its resistance in night and day is RLDR(Night) = 100 KΩ and RLDR(Day) = 500 Ω. The resistance of fixed resistor R1 = 10 KΩ. The LDR and fixed resistor R1 are connected, as shown in Figure 4.110. One end of the LDR is connected to 5 V, and another end is connected to fixed resistor R1. The ground (GND) is connected to another end of resistor R1. The output voltage at the junction of LDR and R1 is feed to analog A0 pin of Arduino UNO board, and a program is written as shown in Figure 4.111, which triggers the relay from Pin 8 of Arduino board depending upon the analog signal received at analog (A0) pin. The LDR-based control circuit works under the following two cases:

Case 1 (Night): The values of various parameters of the control circuit during the night are:

RLDR(Night) = 100 KΩ
R1 = 10 KΩ
V(supply) = 5 V
As per potential divider rule and is given by (equation 3.1)
Vout = V(supply)[R1/(R1 + RLDR(Night))]
or, Vout = 5 V[10 KΩ/(10 KΩ + 100 KΩ)]
or, Vout = 0.45 V

This Vout is feed to analog A0 pin of Arduino board, and the program written for Arduino will not trigger the relay by generating 0 V (binary 0) from Pin 8 of Arduino, which will turn on the bulb (Figure 4.108).

Case 2 (Day): The values of various parameters of the control circuit during the day are:

RLDR(Day) = 500 Ω
R1 = 10 KΩ
V(supply) = 5 V
As per potential divider rule and is given by (equation 3.1)
Vout = V(supply)[R1/(R1 + RLDR(Night))]
or, Vout = 5 V[10 KΩ/(10 KΩ + 500 Ω)]
or, Vout = 4.8 V

This Vout is feed to analog A0 pin of Arduino board, and the program written for Arduino will trigger the relay by generating 5 V (binary 1) from Pin 8 of Arduino, which will turn off the bulb (Figure 4.109).

Description of the Program:

By using the statements (1) to (4), Pin 8 is declared as an output pin, and analog pin A0 is declared as an input pin. The statement (5) initializes the serial communication between the Arduino UNO board and the computer to display the serial monitor at 9,600 baud.

The statement (6) `int step=analogRead(analogInput)` is used to assign 0–1,023 steps to integer “step” depending upon the analog voltage generated by the LDR-based control circuit and received by A0 pin of Arduino UNO.

After experimentation, it is finalized that in the night, the step is less than or equal to 300, and in the day, it is greater than 300.

The statement (7) `if (step <= 300)` is valid for the night, and it will cause statements (8) to (12) to be executed. The statement (8) `digitalWrite(bulb, LOW)` will generate 0 V at Pin 8 of Arduino UNO, and this will not trigger the relay and bulb turn on (Figure 4.108). The statements (9) to (12) display the step and "NIGHT LIGHT ON" message on the serial monitor.

The statement (13) is valid for the day when the step is greater than 300, and it will cause statements (14) to (18) to be executed. The statement (14) `digitalWrite(bulb, HIGH)` will generate 5 V at Pin 8 of Arduino UNO, and this will trigger the relay and bulb turn off (Figure 4.109). The statements (15) to (18) display the step and "DAY LIGHT OFF" message on the serial monitor.

4.14 INTERFACING AND PROGRAMMING OF ARDUINO UNO WITH 4×4 KEYPAD

In this section, we shall discuss the interfacing of 4×4 keypad with Arduino UNO board and programming details. The working and pin description of 4×4 keypad is explained in Section 3.13 of Chapter 3.

Program 4.48

Interface a 4×4 keypad with Arduino UNO board, and write a program to display the pressed key on serial monitor.

Solution

The interfacing of 4×4 keypad with Arduino UNO board is shown in Figure 4.113. The 4×4 keypad Pin 1 (Row 0), 2 (Row 1), 3 (Row 2), 4 (Row 3), 5 (Column 0), 6 (Column 1), 7 (Column 2), and 8 (Column 3) are connected to the Pins 11, 10, 9, 8, 7, 6, 5, and 4 of Arduino board.

An Arduino UNO program to display the pressed key of a 4×4 keypad on the serial monitor for the circuit diagram shown in Figure 4.113 is shown in Figure 4.114.

The screenshot of serial monitor of program shown in Figure 4.114 and interfacing circuit shown in Figure 4.113 to display the pressed key of a 4×4 keypad on the serial monitor is shown in Figure 4.115.

Description of the Program:

The statement (1) `#include <Keypad.h>` is used to include LCD library. The statements (2) and (3) declare the number of rows and columns of 4×4 keypad as four rows and four columns. The statement (4) defines the 4×4 keypad structure, as shown in Figure 4.113. The statement (5) maps the Row 0, Row 1, Row 2, and Row 3 of 4×4 keypad with the Pins 11, 10, 9, and 8, respectively, of Arduino board. The statement (6) maps the Column 0, Column 1, Column 2, and Column 3 of 4×4 keypad with Pins 7, 6, 5,

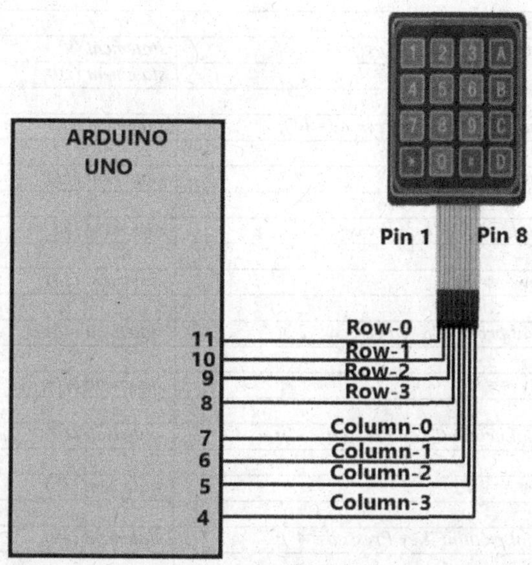

FIGURE 4.113 The interfacing of a 4×4 keypad with Arduino UNO board.

#include <Keypad.h>	*statement (1)*
const byte ROWS = 4; //four rows	*statement (2)*
const byte COLS = 4; //four columns	*statement (3)*
char keys[ROWS][COLS] = *{* *{'1','2','3','A'},* *{'4','5','6','B' },* *{'7','8','9','C'},* *{'*','0','#','D'}* *};*	*statement (4)*
byte rowPins[ROWS] = {11, 10, 9, 8}; *//connect to the row pins of keypad*	*statement (5)*
byte colPins[COLS] = {7, 6, 5, 4}; *//connect to the column pins of keypad*	*statement (6)*
Keypad keypad = Keypad(makeKeymap(keys), rowPins, colPins, ROWS, COLS);	*statement (7)*
void setup()	
{	
Serial.begin(9600);	*statement (8)*
}	

FIGURE 4.114 An Arduino UNO program to display the pressed key of a 4×4 keypad on the serial monitor for the circuit diagram shown in Figure 4.113.

(Continued)

void loop(){	
char key = keypad.getKey();	statement (9)
if (key == '0')	statement (10)
{	
Serial.println("Key Pressed= 0");	statement (11)
}	
if (key == '1')	statement (12)
{	
Serial.println("Key Pressed= 1");	statement (13)
}	
if (key == '2')	statement (14)
{	
Serial.println("Key Pressed= 2");	statement (15)
}	
if (key == '3')	statement (16)
{	
Serial.println("Key Pressed= 3");	statement (17)
}	
if (key == '4')	statement (18)
{	
Serial.println("Key Pressed= 4");	statement (19)
}	
if (key == '5')	statement (20)
{	
Serial.println("Key Pressed= 5");	statement (21)
}	
if (key == '6')	statement (22)
{	
Serial.println("Key Pressed= 6");	statement (23)
}	
if (key == '7')	statement (24)
{	
Serial.println("Key Pressed= 7");	statement (25)
}	
if (key == '8')	statement (26)
{	
Serial.println("Key Pressed= 8");	statement (27)
}	
if (key == '9')	statement (28)
{	
Serial.println("Key Pressed= 9");	statement (29)
}	
if (key == 'A')	statement (30)
{	
Serial.println("Key Pressed= A");	statement (31)
}	
if (key == 'B')	statement (32)
{	
Serial.println("Key Pressed= B");	statement (33)
}	
if (key == 'C')	statement (34)
{	
Serial.println("Key Pressed= C");	statement (35)

FIGURE 4.114 (*CONTINUED*) An Arduino UNO program to display the pressed key of a 4×4 keypad on the serial monitor for the circuit diagram shown in Figure 4.113.

(Continued)

}	
if (key == 'D')	*statement (36)*
{	
Serial.println("Key Pressed= D");	*statement (37)*
}	
if (key == '')*	*statement (38)*
{	
*Serial.println("Key Pressed= *");*	*statement (39)*
}	
if (key == '#')	*statement (40)*
{	
Serial.println("Key Pressed= #");	*statement (41)*
}	
}	

FIGURE 4.114 (*CONTINUED*) An Arduino UNO program to display the pressed key of a 4×4 keypad on the serial monitor for the circuit diagram shown in Figure 4.113.

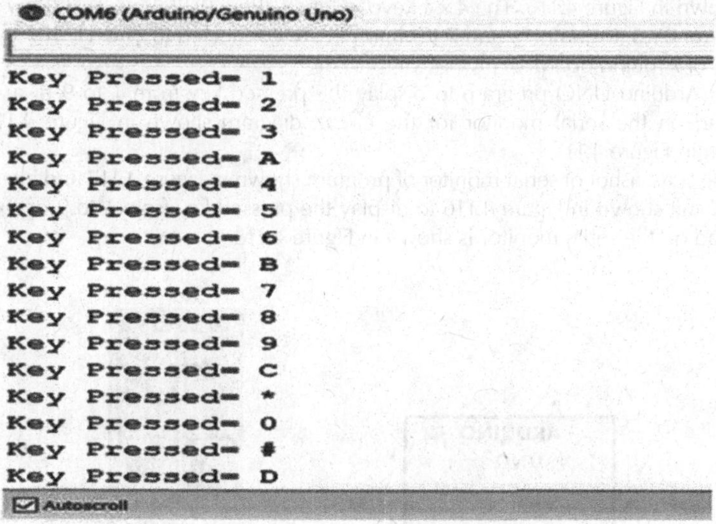

```
COM6 (Arduino/Genuino Uno)

Key Pressed= 1
Key Pressed= 2
Key Pressed= 3
Key Pressed= A
Key Pressed= 4
Key Pressed= 5
Key Pressed= 6
Key Pressed= B
Key Pressed= 7
Key Pressed= 8
Key Pressed= 9
Key Pressed= C
Key Pressed= *
Key Pressed= 0
Key Pressed= #
Key Pressed= D
Autoscroll
```

FIGURE 4.115 The screenshot of serial monitor of program shown in Figure 4.114 and interfacing circuit shown in Figure 4.113 to display the pressed key of a 4×4 keypad on the serial monitor.

and 4, respectively, of Arduino board. The statement (7) creates an object "keypad" and maps it with the rows, columns, and 4×4 keypad structure. The statement (8) initializes the serial communication between the Arduino UNO board and the computer to display the serial monitor at 9,600 baud.

The statement (9) `char key = keypad.getKey()` will use the function `keypad.getKey()`, and the value of whichever key is pressed is assigned to "key". If the pressed key is "0", the statement (10) is true, and the statement (11) `Serial.println("Key Pressed= 0")` will be executed and "Key Pressed= 0" will be displayed on the serial monitor. Similarly, the remaining part of the program from statement (12) to statement (41) will be executed, and the pressed key will be displayed on the serial monitor.

Program 4.49

Let us assume in a specific application we need the keys from (1 to 9) only. Interface a 4×4 keypad with Arduino UNO board, and write a program to display the pressed key from (1 to 9) on the serial monitor.

Solution

The interfacing of a 4×4 keypad with Arduino UNO board such that keys from (1 to 9) only are active is shown in Figure 4.116. All the keys that belong to Column 3 (i.e., A, B, C, and D) and Row 3 (i.e., *, 0, #, and D) are not in use in this application. Therefore, Row 3 and Column 3 pins of the keypad will remain unconnected, as shown in Figure 4.116. The 4×4 keypad Pins 1 (Row 0), 2 (Row 1), 3 (Row 2), 5 (Column 0), 6 (Column 1), and 7 (Column 2) are connected to Pins 11, 10, 9, 7, 6, and 5 of Arduino board.

An Arduino UNO program to display the pressed key from 1 to 9 of a 4×4 Keypad on the serial monitor for the circuit diagram shown in Figure 4.116 is shown in Figure 4.117.

The screenshot of serial monitor of program shown in Figure 4.117 and interfacing circuit shown in Figure 4.116 to display the pressed key from 1 to 9 of a 4×4 Keypad on the serial monitor is shown in Figure 4.118.

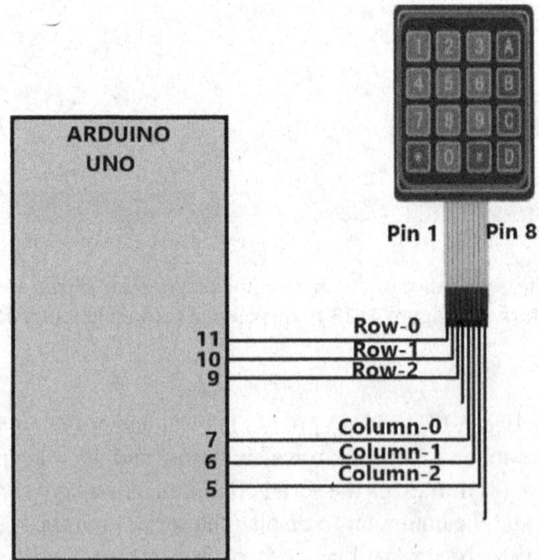

FIGURE 4.116 The interfacing of a 4×4 keypad with Arduino UNO board such that keys from 1 to 9 only are active.

#include <Keypad.h>	statement (1)
const byte ROWS = 3; //three rows	statement (2)
const byte COLS = 3; //three columns	statement (3)

char keys[ROWS][COLS] =
{
{'1','2','3'},
{'4','5','6'},
{'7','8','9'}
};
	statement (4)

byte rowPins[ROWS] = {11, 10, 9}; //connect to the row pins of keypad	statement (5)
byte colPins[COLS] = {7, 6, 5}; //connect to the column pins of keypad	statement (6)
Keypad keypad = Keypad(makeKeymap(keys), rowPins, colPins, ROWS, COLS);	statement (7)
void setup()	
{	
Serial.begin(9600);	statement (8)
}	
void loop()	
{	
char key = keypad.getKey();	statement (9)
if (key == '1')	statement (10)
{	
Serial.println("Key Pressed= 1");	statement (11)
}	
if (key == '2')	statement (12)
{	
Serial.println("Key Pressed= 2");	statement (13)
}	
if (key == '3')	statement (14)
{	
Serial.println("Key Pressed= 3");	statement (15)
}	
if (key == '4')	statement (16)
{	
Serial.println("Key Pressed= 4");	statement (17)
}	
if (key == '5')	statement (18)
{	
Serial.println("Key Pressed= 5");	statement (19)
}	
if (key == '6')	statement (20)
{	
Serial.println("Key Pressed= 6");	statement (21)
}	
if (key == '7')	statement (22)
{	
Serial.println("Key Pressed= 7");	statement (23)
}	
if (key == '8')	statement (24)
{	
Serial.println("Key Pressed= 8");	statement (25)
}	
if (key == '9')	statement (26)
{	
Serial.println("Key Pressed= 9");	statement (27)
}	
}	

FIGURE 4.117 An Arduino UNO program to display the pressed key from 1 to 9 of a 4×4 keypad on the serial monitor for the circuit diagram shown in Figure 4.116.

```
COM6 (Arduino/Genuino Uno)

Key Pressed= 1
Key Pressed= 2
Key Pressed= 3
Key Pressed= 4
Key Pressed= 5
Key Pressed= 6
Key Pressed= 7
Key Pressed= 8
Key Pressed= 9
```

FIGURE 4.118 The screenshot of serial monitor of program shown in Figure 4.117 and interfacing circuit shown in Figure 4.116 to display the pressed key from 1 to 9 of a 4×4 keypad on the serial monitor.

Description of the Program:

The program's description is similar to the description of the program shown in Figure 4.114 except few changes.

The statement (2) and statement (3) declare the number of rows and columns of 4×4 keypad as three rows and three columns. The statement (4) defines the 4×4 keypad as active keys from 1 to 9 only. The statement (5) maps the Row 0, Row 1, and Row 2 of 4×4 keypad with Pins 11, 10, and 9, respectively, of Arduino board. The statement (6) maps the Column 0, Column 1, and Column 2 of 4×4 keypad with Pins 7, 6, and 5, respectively, of Arduino board.

Program 4.50

Interface a 4×4 keypad and a LED with Arduino UNO board, and write a program to turn on the LED if key 1 is pressed and turn off the LED if key 2 is pressed, and display the message on the serial monitor.

Solution

The interfacing of a 4×4 keypad and a LED with Arduino UNO board is shown in Figure 4.119. We need only key 1 and key 2 of 4×4 keypad; therefore, Row 1, Column 0, and Column 1 are required. The 4×4 keypad Pins 1 (Row 0), 5 (Column 0), and 6 (Column 1) are connected to the Pins 11, 7, and 6 of Arduino board. The anode of LED is connected to Pin 2 of Arduino UNO board through a 250 Ω resistor, and the cathode is connected to the GND (ground) pin of Arduino board.

An Arduino UNO program to turn on the LED if key 1 is pressed and turn off the LED if key 2 is pressed and display the status of LED and pressed key on the serial monitor for the circuit diagram shown in Figure 4.119 is shown in Figure 4.120.

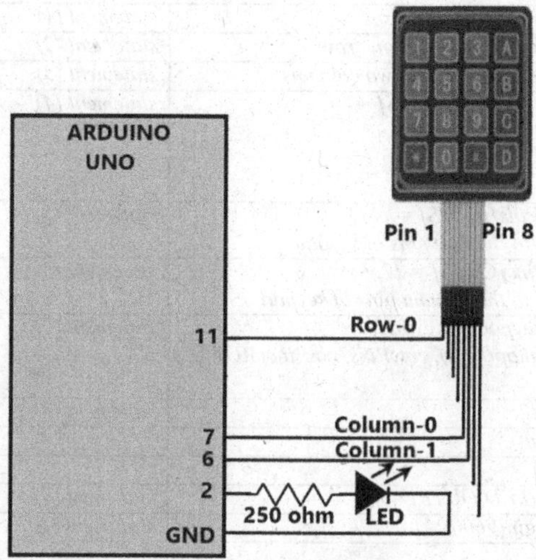

FIGURE 4.119 The interfacing of a 4×4 keypad and a LED with Arduino UNO board.

The screenshot of serial monitor of program shown in Figure 4.120 and interfacing circuit shown in Figure 4.119 to display the status of LED and pressed key on the serial monitor is shown in Figure 4.121.

Description of the Program:

The program's description is similar to the description of the program shown in Figure 4.114 except few changes.

The statement (2) and statement (3) declare the number of rows and columns of 4×4 keypad as one row and two columns. The statement (4) defines the 4×4 keypad as active keys 1 and 2 only. The statement (5) maps the Row 0 of 4×4 keypad with Pin 11 of Arduino board. The statement (6) maps the Column 0 and Column 1 of 4×4 keypad with Pins 7 and 6, respectively, of Arduino board. The statement (8) will give Pin 2 of the Arduino board the name as "LED". The statement (9) will declare Pin 2 as an output pin. Due to the statements (12), (13), (14), and (15), the LED will turn on when key 1 is pressed, and an appropriate message will be displayed on the serial monitor. Due to the statements (16), (17), (18), and (19), the LED will turn off when key 2 is pressed, and an appropriate message will be displayed on the serial monitor.

4.15 INTERFACING AND PROGRAMMING OF ARDUINO UNO WITH OPTICAL SENSOR

In this section, we shall discuss the interfacing of optical sensor with Arduino UNO board and programming details. An optical sensor is a device that converts the light rays into electrical signals. The working and pin description of optical sensor is explained in Section 3.14 of Chapter 3.

#include <Keypad.h>	statement (1)
const byte ROWS = 1; //one row	statement (2)
const byte COLS = 2; //two columns	statement (3)

char keys[ROWS][COLS] =
{
{'1','2'}
};
	statement (4)

byte rowPins[ROWS] = {11}; //connect to the row pins of keypad	statement (5)
byte colPins[COLS] = {7, 6}; //connect to the column pins of keypad	statement (6)
Keypad keypad = Keypad(makeKeymap(keys), rowPins, colPins, ROWS, COLS);	statement (7)
int LED=2;	statement (8)
void setup()	
{	
pinMode(LED,OUTPUT);	statement (9)
Serial.begin(9600);	statement (10)
}	
void loop()	
{	
char key = keypad.getKey();	statement (11)
if (key == '1')	statement (12)
{	
digitalWrite(LED, HIGH);	statement (13)
Serial.println("Key Pressed= 1");	statement (14)
Serial.println("LED ON");	statement (15)
}	
if (key == '2')	statement (16)
{	
digitalWrite(LED, LOW);	statement (17)
Serial.println("Key Pressed= 2");	statement (18)
Serial.println("LED OFF");	statement (19)
}	
}	

FIGURE 4.120 An Arduino UNO program to turn on the LED if key 1 is pressed and turn off the LED if key 2 is pressed and display the status of LED and pressed key on the serial monitor for the circuit diagram shown in Figure 4.119.

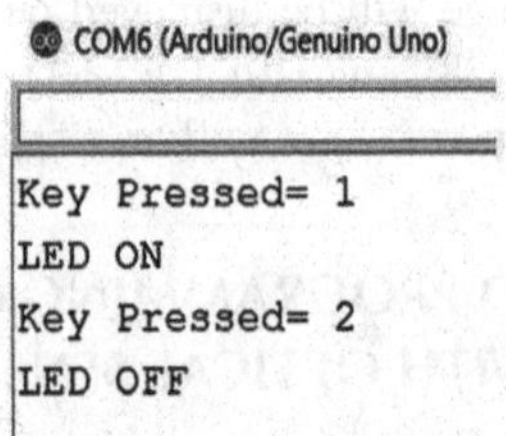

FIGURE 4.121 The screenshot of serial monitor of program shown in Figure 4.120 and interfacing circuit shown in Figure 4.119 to display the status of LED and pressed key on the serial monitor.

Program 4.51 (a)

Interface a retro-reflective optical sensor module with Arduino UNO board, and write a program to determine the output generated by the optical sensor module in the presence and absence of obstacle and display the information on the serial monitor.

Solution

The interfacing of a retro-reflective optical sensor module with Arduino UNO board is shown in Figure 4.122. The Vcc pin, ground pin, and output pin of optical sensor module are connected to 5 V, GND (ground), and Pin 2 of Arduino board.

An Arduino UNO program to find out the output generated by the optical sensor module in the presence and absence of obstacle and also display the information on the serial monitor for the circuit diagram shown in Figure 4.122 is shown in Figure 4.123.

The screenshot of serial monitor of program shown in Figure 4.123 and interfacing circuit shown in Figure 4.122 to display the output generated by the optical sensor module in the presence and absence of obstacle on the serial monitor is shown in Figure 4.124.

Description of the Program:

Since the optical sensor's output pin is connected to Pin 2 of the Arduino board and the sensor's output voltage level will enter inside Arduino from Pin 2, it has to be an input pin. Using the statements (1) and (2), Pin 2 is declared as an input pin. The statement (3) initializes the serial communication between the Arduino UNO board and the computer to display the serial monitor at 9,600 baud.

The statement (4) will assign the sensor's output voltage level to an integer variable "opticalSensorLevel".

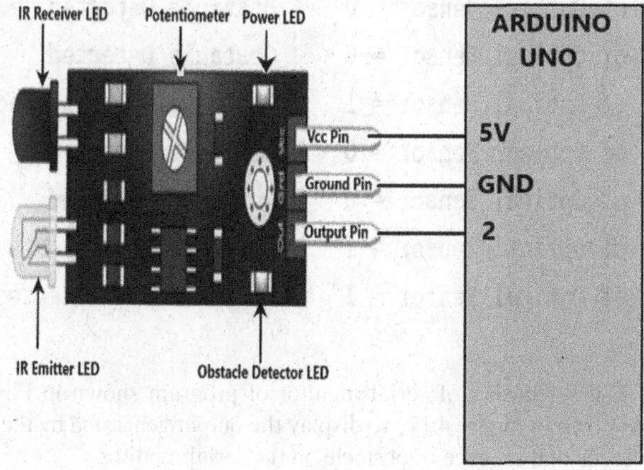

FIGURE 4.122 The interfacing of a retro-reflective optical sensor module with Arduino UNO board.

int opticalSensorValue=2;	*statement (1)*
void setup()	
{	
pinMode(opticalSensorValue,INPUT);	*statement (2)*
Serial.begin(9600);	*statement (3)*
}	
void loop()	
{	
int opticalSensorLevel = digitalRead(opticalSensorValue);	*statement (4)*
Serial.print ("Output of Optical Sensor = ");	*statement (5)*
Serial.print(opticalSensorLevel);	*statement (6)*
if(opticalSensorLevel==0)	*statement (7)*
{	
Serial.println (" Obstacle Detected");	*statement (8)*
}	
else	*statement (9)*
{	
Serial.println (" Obstacle Not Detected");	*statement (10)*
}	
delay(5000);	*statement (11)*
}	

FIGURE 4.123 An Arduino UNO program to find out the output generated by the optical sensor module in the presence and absence of obstacle and also display the information on the serial monitor for the circuit diagram shown in Figure 4.122.

COM6 (Arduino/Genuino Uno)

```
Output of Optical Sensor = 1    Obstacle Not Detected
Output of Optical Sensor = 1    Obstacle Not Detected
Output of Optical Sensor = 0    Obstacle Detected
Output of Optical Sensor = 0    Obstacle Detected
Output of Optical Sensor = 1    Obstacle Not Detected
Output of Optical Sensor = 0    Obstacle Detected
Output of Optical Sensor = 0    Obstacle Detected
Output of Optical Sensor = 1    Obstacle Not Detected
Output of Optical Sensor = 1    Obstacle Not Detected
```

FIGURE 4.124 The screenshot of serial monitor of program shown in Figure 4.123 and interfacing circuit shown in Figure 4.122 to display the output generated by the optical sensor module in the presence and absence of obstacle on the serial monitor.

If the sensor's output voltage level is Logic 0, then statements (5) to (8) display the voltage level 0 and "Obstacle Not Detected". If the output voltage level of the sensor is Logic 1, then statements (5) to (6) and statements (9) to (10) display the voltage level 1 and "Obstacle Detected". The statement (11) will generate a delay of 5 seconds.

Program 4.51 (b)

Interface a retro-reflective optical sensor module, relay module, and bulb with Arduino UNO board, and write a program to turn on the bulb when optical sensor module detects an obstacle.

Solution

The interfacing of 5 V relay board, 220 V AC-operated bulb, and optical sensor with Arduino UNO board is shown in Figure 4.125. The 220 V AC-operated bulb is connected to NC (normally close) and COM (also called as pole) pins of relay. The 5 V and GND (ground) pins of relay are connected to 5 V and GND (ground) pins of Arduino board. The IN pin of relay is connected to Pin 8 of Arduino board. The Vcc pin, ground pin, and output pin of optical sensor module are connected to 5 V, GND (ground), and Pin 2 of Arduino board.

The schematic of a 220 V AC-operated bulb connected to COM and NC (normally close) terminals of relay is shown in Figure 4.126. The relay is not

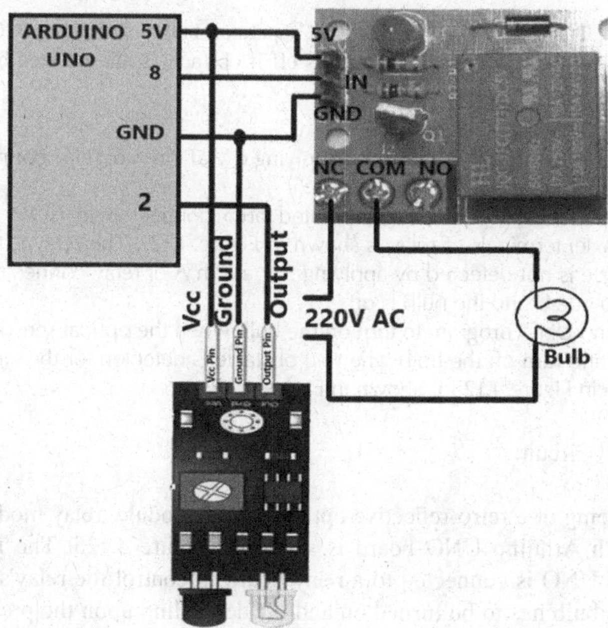

FIGURE 4.125 The interfacing of a retro-reflective optical sensor module, relay module, and AC-operated bulb with Arduino UNO board.

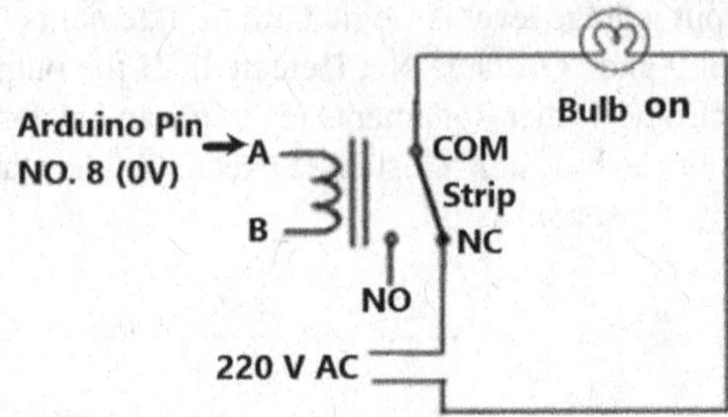

FIGURE 4.126 The schematic of a 220 V AC-operated bulb connected to COM and NC (normally close) terminals of relay. The bulb is on if obstacle is detected by not triggering the relay.

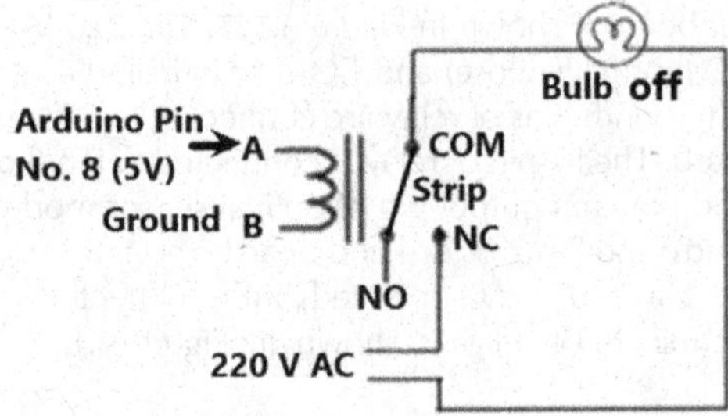

FIGURE 4.127 The schematic of a 220 V AC-operated bulb connected to COM and NC (normally close) terminals of relay. The bulb is ôff if obstacle is not detected by triggering the relay.

triggered when obstacle is detected by applying 0 V at Pin A of relay connected to Pin 8 of Arduino board and the bulb is on.

The schematic of a 220 V AC-operated bulb connected to COM and NC (normally close) terminals of relay is shown in Figure 4.127. The relay is triggered when obstacle is not detected by applying 5 V at Pin A of relay connected to Pin 8 of Arduino board and the bulb is off.

An Arduino UNO program to turn on the bulb when the optical sensor detects an obstacle and turn off the bulb when no obstacle is detected for the circuit diagram shown in Figure 4.125 is shown in Figure 4.128.

Working of the Circuit:

The interfacing of a retro-reflective optical sensor module, relay module, and bulb with Arduino UNO board is shown in Figure 4.125. The Pin 8 of Arduino UNO is connected to a relay board to control the relay's triggering. The bulb has to be turned on and off depending upon the presence or absence of the obstacle. The program is developed such that if there is an obstacle in front of the optical sensor, then Pin 8 of Arduino board will generate Logic 0, which does not trigger the relay and the bulb is turned on. If there is no obstacle in front of the optical sensor, then Pin 8 of the Arduino board will generate Logic 1, which will trigger the relay, and the bulb is turned off. Turning on and off of the bulb is done using relay and sensing of the obstacle done by the optical sensor. When the relay does not trigger, the connection between COM and NC terminals of the relay is complete, and this will cause the bulb to be turned on (see Figure 4.126). If the relay is triggered, then the connection between COM and NO terminals of the relay is complete, and this will cause the bulb to be turned off (see Figure 4.127).

Description of the Program:

Using the statements (1) and (3), the Pin 2 Arduino board is declared as an input pin and the name assigned to Pin 2 is "opticalSensorValue". Using the statements (2) and (4), Pin 8 Arduino board is declared as an output pin and the name assigned to Pin 8 is "relay". The statement (5) will assign the optical sensor's output voltage level to an integer variable "opticalSensorLevel". Suppose there is an obstacle in front of the optical sensor. In that case, the sensor output generates 0 V and statement (6) `if(opticalSensorLevel==0)` is true and statement (7) `digitalWrite(relay, LOW)` will be executed to cause relay not to trigger, and the bulb will turn on. Suppose there is no obstacle in front of the optical sensor. In that case, the sensor output generates 5 V (Logic 1)

int opticalSensorValue=2;	*statement (1)*
int relay=8;	*statement (2)*
void setup()	
{	
pinMode(opticalSensorValue,INPUT);	*statement (3)*
pinMode(relay,OUTPUT);	*statement (4)*
}	
void loop()	
{	
int opticalSensorLevel = digitalRead(opticalSensorValue);	*statement (5)*
if(opticalSensorLevel==0)	*statement (6)*
{	
digitalWrite(relay,LOW); //obstacle detected, bulb on	*statement (7)*
}	
if(opticalSensorLevel==1)	*statement (8)*
{	
digitalWrite(relay,HIGH); //obstacle does not detected, bulb off	*statement (9)*
}	
delay(1000);	*statement (10)*
}	

FIGURE 4.128 An Arduino UNO program to turn on the bulb when the optical sensor detects an obstacle and turn off the bulb when no obstacle is detected for the circuit diagram shown in Figure 4.125.

and statement (8) `if(opticalSensorLevel==1)` is true and statement (9) `digitalWrite(relay, HIGH)` will be executed to cause the relay to trigger, and the bulb will turn off. The statement (10) will generate a delay of 1 second.

4.16 INTERFACING OF CAPACITIVE TOUCH SENSOR WITH ARDUINO UNO

This section shall discuss the interfacing of the capacitive touch sensor with Arduino UNO board and programming details. A capacitive touch sensor works on the principle of capacitive effect, which can detect the physical touch. The working and pin description of capacitive touch sensor is explained in Section 3.15 of Chapter 3.

Program 4.52

Interface a capacitive touch sensor module with Arduino UNO board, and write a program to determine the touch sensor module's output when we touch it and display the information on the serial monitor.

Solution

The interfacing of a touch sensor module with Arduino UNO board is shown in Figure 4.129. The VCC, GND (ground), and SIG (output) pins of touch sensor module are connected to 5 V, GND (ground), and 11 pins of Arduino board.

An Arduino UNO program to find out the output generated by the capacitive touch sensor module when we touch it and display the information on the serial monitor for the circuit diagram shown in Figure 4.129 is shown in Figure 4.130.

The screenshot of serial monitor of program shown in Figure 4.130 and interfacing circuit shown in Figure 4.129 to display the status and output generated by the capacitive touch sensor module when we touch/not touch it on the serial monitor is shown in Figure 4.131.

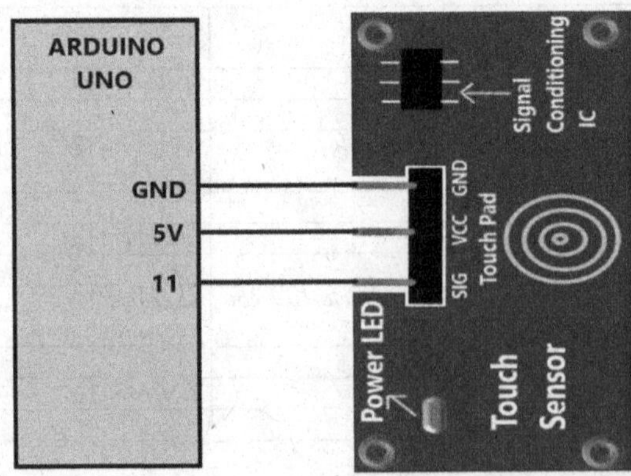

FIGURE 4.129 The interfacing of a capacitive touch sensor module with Arduino UNO board.

int touchSensor=11;	*statement (1)*
void setup()	
{	
pinMode(touchSensor,INPUT);	*statement (2)*
Serial.begin(9600);	*statement (3)*
}	
void loop()	
{	
int touchValue=digitalRead(touchSensor);	*statement (4)*
Serial.print("Capacitive touch sensor value=");	*statement (5)*
Serial.print(touchValue);	*statement (6)*
if (touchValue==0)	*statement (7)*
{	
Serial.println(" Touch Not Detected");	*statement (8)*
}	
if (touchValue==1)	*statement (9)*
{	
Serial.println(" Touch Detected");	*statement (10)*
}	
delay(3000);	*statement (11)*
}	

FIGURE 4.130 An Arduino UNO program to find out the output generated by the capacitive touch sensor module when we touch it and display the information on the serial monitor for the circuit diagram shown in Figure 4.129.

```
COM6 (Arduino/Genuino Uno)

Capacitive touch sensor value=0     Touch Not Detected
Capacitive touch sensor value=0     Touch Not Detected
Capacitive touch sensor value=1     Touch Detected
Capacitive touch sensor value=1     Touch Detected
Capacitive touch sensor value=1     Touch Detected
Capacitive touch sensor value=0     Touch Not Detected
Capacitive touch sensor value=0     Touch Not Detected
```

FIGURE 4.131 The screenshot of serial monitor of program shown in Figure 4.130 and interfacing circuit shown in Figure 4.129 to display the status and output generated by the capacitive touch sensor module when we touch/not touch it on the serial monitor.

Description of the Program:

The output pin of capacitive touch sensor is connected to Pin 11 of the Arduino board. The sensor's output voltage level will enter inside Arduino from Pin 11; therefore, it has to be an input pin. Using the statements (1) and (2), Pin 11 is declared an input pin. The statement (3) initializes the serial communication between the Arduino UNO board and the computer to display the serial monitor at 9,600 baud. The statement (4) will assign the sensor's output voltage level to an integer variable "touchValue". Let's touch the touchpad of the sensor. The digital output pin of the capacitive touch sensor module generates 5 V (Logic 1); otherwise, it generates 0 V (Logic 0), and this value is stored in variable "touchValue". If the sensor is not touched, then the sensor's output pin generates 0 V and Pin 11 of Arduino receives Logic 0, and it will cause statements (7) and (8) to execute. If the sensor is touched, then the sensor's output pin generates 5 V, and Pin 11 of Arduino receives Logic 1, and it will cause statements (9) and (10) to execute. The statement (11) will generate a delay of 3 seconds.

Program 4.53

Interface a capacitive touch sensor module, relay module, and bulb with Arduino UNO board, and write a program to turn off the bulb when touch sensor module is touched.

Solution

The interfacing of 5 V relay board, 220 V AC-operated bulb, and touch sensor with Arduino UNO board is shown in Figure 4.132. The 220 V AC-operated bulb is connected to NC (normally close) and COM (also called as pole) pins of relay. The 5 V and GND (ground) pins of relay are connected to 5 V and GND (ground) pins of Arduino board. The IN pin of relay is connected to Pin 13 of Arduino board. The

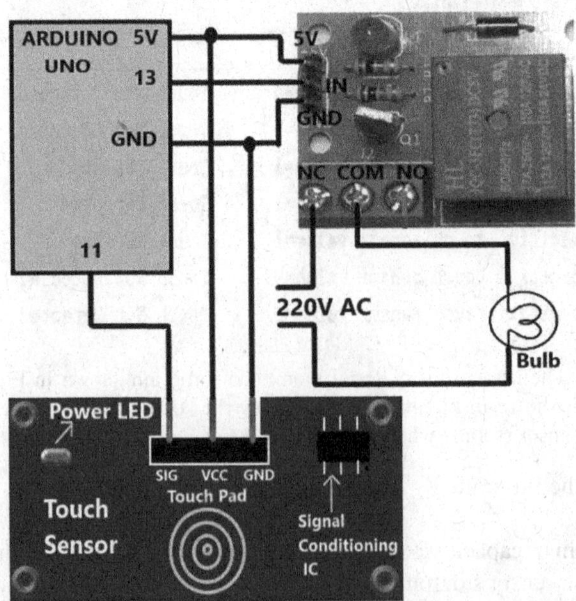

FIGURE 4.132 The interfacing of a capacitive touch sensor module, relay module, and bulb with Arduino UNO board.

VCC, GND (ground), and SIG (output) pins of touch sensor module are connected to 5 V, GND (ground), and Pin 11 of Arduino board.

The schematic of a 220 V AC-operated bulb connected to COM and NC (normally close) terminals of relay is shown in Figure 4.133. The relay is not triggered if touch sensor is not touched by applying 0 V at pin A of relay connected to Pin 13 of Arduino board and the bulb is on.

The schematic of a 220 V AC-operated bulb connected to COM and NC (normally close) terminals of relay is shown in Figure 4.134. The relay is triggered if touch sensor is touched applying 5 V at Pin A of relay connected to Pin 13 of Arduino board and the bulb is off.

An Arduino UNO program to turn off the bulb when touch sensor module is touched for the circuit diagram shown in Figure 4.132 is shown in Figure 4.135.

Working of the Circuit:

The interfacing of a capacitive touch sensor module, relay module, and bulb with Arduino UNO board is shown in Figure 4.132. The Pin 13 of Arduino UNO is connected to the relay board to control the relay's

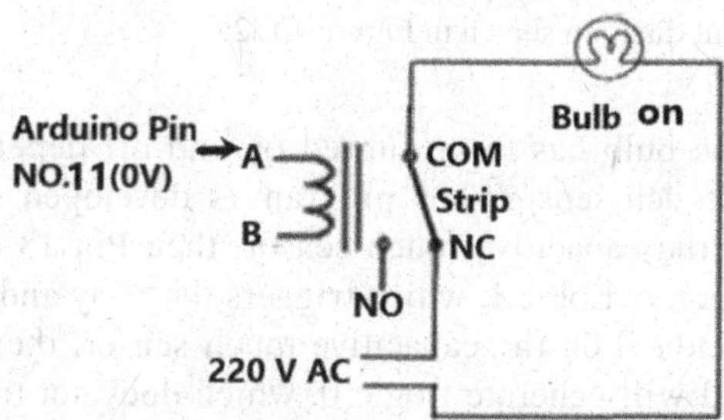

FIGURE 4.133 The schematic of a 220 V AC-operated bulb connected to COM and NC (normally close) terminals of relay. The bulb is on if touch sensor is not touched by not triggering the relay.

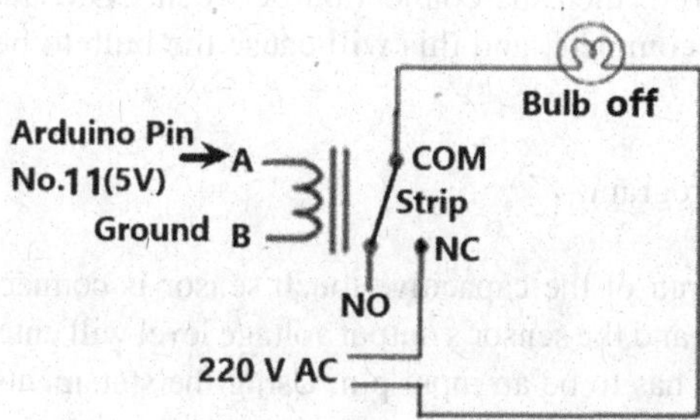

FIGURE 4.134 The schematic of a 220 V AC-operated bulb connected to COM and NC (normally close) terminals of relay. The bulb is off if touch sensor is touched by triggering the relay.

int touchSensor=11;	*statement (1)*
int relay=13;	*statement (2)*
void setup()	
{	
pinMode(touchSensor,INPUT);	*statement (3)*
pinMode(relay,OUTPUT);	*statement (4)*
}	
void loop()	
{	
int touchValue=digitalRead(touchSensor);	*statement (5)*
if (touchValue==0)	*statement (6)*
{	
digitalWrite(relay, LOW);//touch not detected, bulb is on	*statement (7)*
}	
if (touchValue==1)	*statement (8)*
{	
digitalWrite(relay, HIGH);//touch detected, bulb is off	*statement (9)*
}	
delay(3000);	*statement (10)*
}	

FIGURE 4.135 An Arduino UNO program to turn off the bulb when touch sensor module is touched for the circuit diagram shown in Figure 4.132.

triggering. The bulb has to be turned on and off depending upon the touch on the touch sensor. The program is developed so that if there is a touch on the capacitive touch sensor, then Pin 13 of the Arduino board will generate Logic 1, which triggers the relay and the bulb is off. If there is no touch on the capacitive touch sensor, then Pin 13 of the Arduino board will generate Logic 0, which does not trigger the relay, and the bulb is turned on.

Turning on and off the bulb is done using the relay, and the capacitive touch sensor module does touch sensing. When the relay does not trigger, the connection between COM and NC terminals of the relay is complete, and this will cause the bulb to be turned on (see Figure 4.133). If the relay is triggered, then the connection between COM and NO terminals of the relay is complete, and this will cause the bulb to be turned off (see Figure 4.134).

Description of the Program:

Since the output pin of the capacitive touch sensor is connected to Pin 11 of Arduino board and the sensor's output voltage level will enter inside Arduino from Pin 11, it has to be an input pin. Using the statements (1) and (3), Pin 11 of Arduino board is declared as an input pin and the name assigned to Pin 11 is "touch sensor". Using the statements (2) and (4), Pin 13 Arduino board is declared as an output pin and the name assigned to Pin 13 is "relay". The statement (5) will assign the sensor's output voltage level to an integer variable "touchValue". Let's touch the touchpad of the sensor. Digital output pin of the capacitive touch sensor module generates 5 V (Logic 1); otherwise, it generates 0 V (Logic 0), and this value is stored in variable "touchValue". If the sensor is not touched, then sensor's output pin generates 0 V and Pin 11 of Arduino receives Logic 0, and it will cause statement (7) to execute and Pin 13 of Arduino will generate 0 V and the relay will not trigger, and this will make bulb turn on. If the sensor is touched, then sensor's output pin generates 5 V and Pin 11 of Arduino receives Logic 1, and it will cause statement (9) to execute and Pin 13 of Arduino will generate 5 V, and the relay will trigger, and this will make bulb turn off. The statement (10) will generate a delay of 3 seconds.

4.17 INTERFACING AND PROGRAMMING OF ARDUINO UNO WITH SMOKE DETECTOR SENSOR

This section shall discuss the interfacing of the smoke detector sensor (gas sensor) with Arduino UNO board and programming details. Gas sensor module (MQ2) can be used for sensing LPG, smoke, alcohol, propane, hydrogen, methane, and carbon monoxide concentrations in the air. The MQ2 gas sensor is used for the description of interfacing circuits and programs. The working principle of MQ2 is explained in Section 3.16 of Chapter 3.

Program 4.54

Interface a smoke detector sensor module with Arduino UNO board, and write a program to determine the output generated by the smoke detector sensor module when smoke is detected and display the information on the serial monitor.

Solution

The interfacing of a smoke detector sensor module (MQ-2) with Arduino UNO board is shown in Figure 4.136. The Vcc, GND (ground), and AO (analog output) pins of smoke detector sensor module are connected to 5 V, GND (ground), and A0 pins of Arduino board.

An Arduino UNO program to detect the smoke by using the smoke detector sensor module (MQ-2) and display the information on the serial monitor for the circuit diagram shown in Figure 4.136 is shown in Figure 4.137.

The screenshot of serial monitor of program shown in Figure 4.137 and interfacing circuit shown in Figure 4.136 to display the status of smoke detection and sensor output on the serial monitor is shown in Figure 4.138.

Description of the Program:

Since the analog output pin of smoke detector sensor is connected to analog pin A0 Arduino board and the sensor's output voltage level will enter inside Arduino from the A0 pin, it has to be an input pin. Using the statements (1) and (2), the A0 pin is given a name "MQ2analogOutput" and declared as

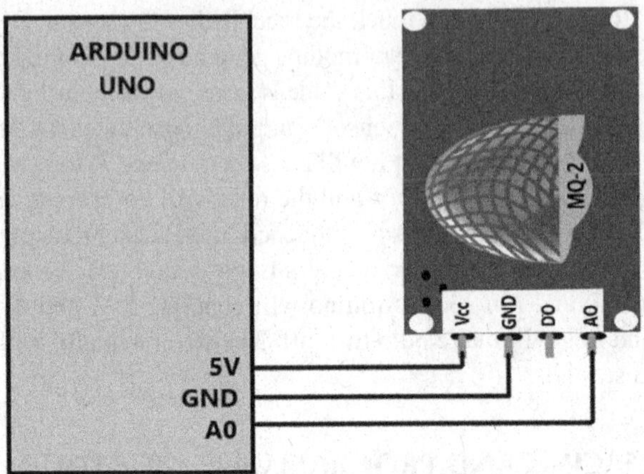

FIGURE 4.136 The interfacing of a smoke detector sensor module with Arduino UNO board.

int MQ2analogOutput=A0;	*statement (1)*
void setup()	
{	
pinMode(MQ2analogOutput,INPUT);	*statement (2)*
Serial.begin(9600);	*statement (3)*
Serial.println("Gas sensor is warming");	*statement (4)*
delay(20000);	*statement (5)*
}	
void loop()	
{	
int sensorValue= analogRead(MQ2analogOutput);	*statement (6)*
Serial.print("Sensor Value= ");	*statement (7)*
Serial.print(sensorValue);	*statement (8)*
if(sensorValue > 300)	*statement (9)*
{	
Serial.println(" Smoke detected");	*statement (10)*
}	
else	*statement (11)*
{	
Serial.println(" No Smoke");	*statement (12)*
}	
delay(5000);	*statement (13)*
}	

FIGURE 4.137 An Arduino UNO program to detect the smoke by using the smoke detector sensor module (MQ-2) and display the information on the serial monitor for the circuit diagram shown in Figure 4.136.

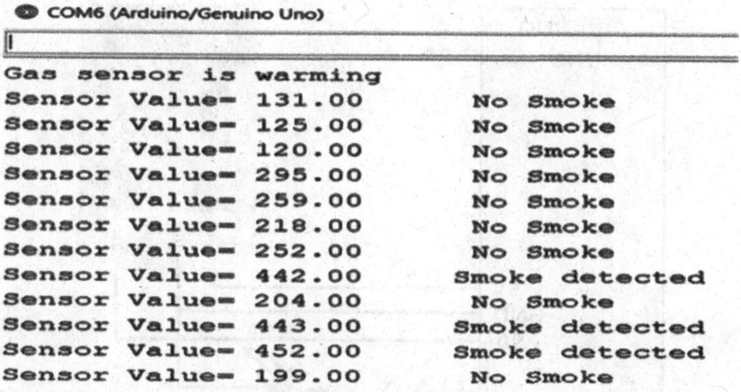

```
COM6 (Arduino/Genuino Uno)

Gas sensor is warming
Sensor Value= 131.00        No Smoke
Sensor Value= 125.00        No Smoke
Sensor Value= 120.00        No Smoke
Sensor Value= 295.00        No Smoke
Sensor Value= 259.00        No Smoke
Sensor Value= 218.00        No Smoke
Sensor Value= 252.00        No Smoke
Sensor Value= 442.00      Smoke detected
Sensor Value= 204.00        No Smoke
Sensor Value= 443.00      Smoke detected
Sensor Value= 452.00      Smoke detected
Sensor Value= 199.00        No Smoke
```

FIGURE 4.138 The screenshot of serial monitor of program shown in Figure 4.137 and interfacing circuit shown in Figure 4.136 to display the status of smoke detection and sensor output on the serial monitor.

an input pin, respectively. The statement (3) initializes the serial communication between the Arduino UNO board and the computer to display the serial monitor at 9,600 baud. To initiate the gas/smoke detection, the sensing element must be warm up for 20 seconds; therefore, we have written statement (5), which generates 20-second delay. The statement (4) displays "Gas sensor is warming" message on the serial monitor.

The statement (6) `int sensorValue=analogRead(MQ2analogoutput)` is used to assign 0–1,023 steps to integer "sensorValue" for the analog output generated by the smoke detector, which in turn depends on the intensity of smoke around the smoke sensor.

The statements (7) and (8) will display the smoke's "sensorValue" on the serial monitor.

With the experimentation, we concluded that if the "senseValue" of the smoke is greater than 300, then "Smoke detected" message is to be displayed on the serial monitor. The values can be different from 300 depending upon the particular requirement. This will be achieved by statements (9) and (10). If the "senseValue" of the smoke is less than or equal to 300, then "No Smoke" message will be displayed on the serial monitor by using statements (11) and (12). The statement (11) will generate a delay of 5 seconds.

Program 4.55

Interface a smoke detector sensor module and a LED with Arduino UNO board, and write a program to turn on the LED if the smoke detector sensor module detects smoke.

Solution

The interfacing of a smoke detector sensor module (MQ-2) and LED with Arduino UNO board is shown in Figure 4.139. The Vcc, GND (ground), and AO (analog

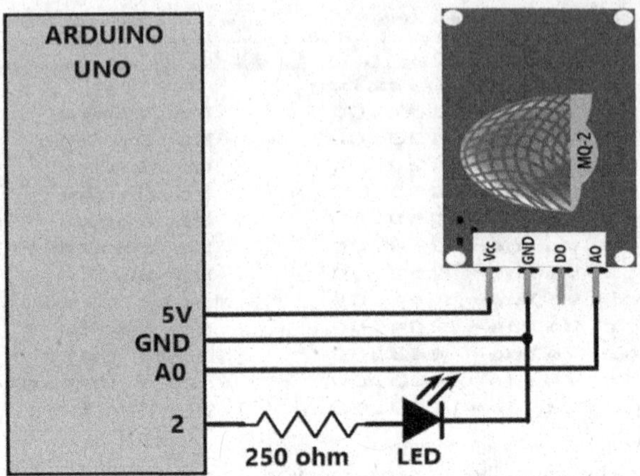

FIGURE 4.139 The interfacing of a smoke detector sensor module and a LED with Arduino UNO board.

int MQ2analogOutput=A0;	*statement (1)*
int LED=2;	*statement (2)*
void setup()	
{	
pinMode(MQ2analogOutput,INPUT);	*statement (3)*
pinMode(LED,OUTPUT);	*statement (4)*
Serial.println("Gas sensor is warming");	*statement (5)*
delay(20000);	*statement (6)*
}	
void loop()	
{	
int sensorValue= analogRead(MQ2analogOutput);	*statement (7)*
if(sensorValue > 300)	*statement (8)*
{	
digitalWrite(LED,HIGH);	*statement (9)*
}	
else	*statement (10)*
{	
digitalWrite(LED,LOW);	*statement (11)*
}	
delay(5000);	*statement (12)*
}	

FIGURE 4.140 An Arduino UNO program to turn on the LED if the smoke is detected by the smoke sensor module (MQ-2) for the circuit diagram shown in Figure 4.139.

output) pins of smoke detector sensor module are connected to 5 V, GND (ground), and A0 pins of Arduino board. The anode of LED is connected to Pin 2 of Arduino UNO board through a 250 Ω resistor, and the cathode is connected to the GND (ground) pin of Arduino board.

An Arduino UNO program to turn on the LED if the smoke is detected by the smoke sensor module (MQ-2) for the circuit diagram shown in Figure 4.139 is shown in Figure 4.140.

Description of the Program:

The program is almost similar to the program shown in Figure 4.137. The program shown in Figure 1.140 will turn on the LED if the smoke's "sensorValue" is greater than 300.

4.18 INTERFACING AND PROGRAMMING OF ARDUINO UNO WITH RAIN DETECTOR SENSOR (FC-07)

This section shall discuss the interfacing of rain detector sensor with Arduino UNO board and programming details. The rain detector sensor module is used for the detection of rain. The FC-07 rain detector sensor is used for the description of interfacing circuits and programs. The working and pin description of FC-07 rain detector sensor is explained in Section 3.17 of Chapter 3.

Program 4.56

Interface a rain detector sensor module (FC-07) with Arduino UNO board, and write a program to determine the output generated by the rain detector sensor module when water is detected and display the information on the serial monitor.

Solution

The interfacing of a rain detector sensor module with Arduino UNO board is shown in Figure 4.141. The VCC, GND (ground), and AO (analog output) pins of the rain detector sensor module are connected to 5 V, GND (ground), and A5 pins of Arduino board.

An Arduino UNO program to find out the output generated by the rain detector sensor module when water is detected and display the steps of internal analog-to-digital converter on the serial monitor for the circuit diagram shown in Figure 4.141 is shown in Figure 4.142.

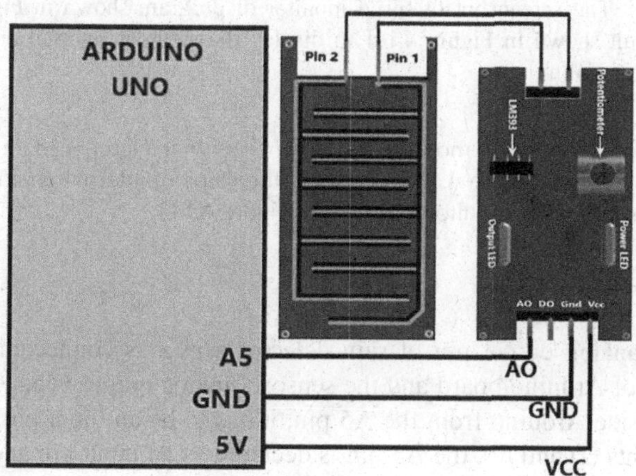

FIGURE 4.141 The interfacing of a rain detector sensor module with Arduino UNO board.

int rainSensorAnalogOutput = A5;	*statement (1)*
void setup()	
{	
pinMode(rainSensorAnalogOutput,INPUT);	*statement (2)*
Serial.begin(9600);	*statement (3)*
}	
void loop()	
{	
int step=analogRead(rainSensorAnalogOutput);	*statement (4)*
Serial.print("Step of ADC=");	*statement (5)*
Serial.println(step);	*statement (6)*
delay(5000);	*statement (7)*
}	

FIGURE 4.142 An Arduino UNO program to find out the output generated by the rain detector sensor module when water is detected and display the steps of internal analog-to-digital converter on the serial monitor for the circuit diagram shown in Figure 4.141.

```
COM6 (Arduino/Genuino Uno)

Step of ADC=1022
Step of ADC=1022
Step of ADC=1022
Step of ADC=1023
Step of ADC=1022
Step of ADC=491
Step of ADC=438
Step of ADC=303
Step of ADC=262
Step of ADC=732
Step of ADC=1010
Step of ADC=278
Step of ADC=251
Step of ADC=377
```

FIGURE 4.143 The screenshot of serial monitor of program shown in Figure 4.142 and interfacing circuit shown in Figure 4.141 to display the steps of internal analog-to-digital converter on the serial monitor.

The screenshot of serial monitor of program shown in Figure 4.142 and interfacing circuit shown in Figure 4.141 to display the steps of internal analog-to-digital converter on the serial monitor is shown in Figure 4.143.

Description of the Program:

Since the analog output pin of rain detector sensor is connected to analog Pin A5 of Arduino board and the sensor's analog output voltage level will enter inside Arduino from the A5 pin, it has to be an input pin. Using the statements (1) and (2), the A5 pin is declared as an input pin and the name assigned to it is "rainSensorAnalogOutput". The statement (3) initializes the serial communication between the Arduino UNO board and the computer to display the serial monitor at 9,600 baud. The statement (4) `int step=analogRead(rainSensorAnalogOutput)` is used to assign 0–1,023 steps to integer "step" depending upon the analog output generated by the rain sensor, which in turn depends upon the water falling on rain board. If there is no water, then the step's value will be more, and it decreases with the increasing quantity of water falling on the rain board. The statements (5) and (6) will display the "step" of the analog input on the serial monitor. The statement (7) will generate a delay of 5 seconds.

Program 4.57

Interface a rain detector sensor module (FC-07) and a LED with Arduino UNO board, and write a program to turn on the LED if it is raining heavily and displaying the information on serial monitor.

Solution

The interfacing of a rain detector sensor module (FC-07) and LED with Arduino UNO board is shown in Figure 4.144. The VCC, GND (ground), and AO (analog output) pins of rain detector sensor module are connected to 5 V, GND (ground), and A5 pins of Arduino board. The anode of LED is connected to Pin 2 of Arduino UNO board through a 250 Ω resistor, and the cathode is connected to the GND (ground) pin of Arduino board.

An Arduino UNO program to turn on the LED if the rain is detected by the rain detector sensor module and display the status of LED and rain condition on the serial monitor for the circuit diagram shown in Figure 4.144 is shown in Figure 4.145.

The screenshot of serial monitor of program shown in Figure 4.145 and interfacing circuit shown in Figure 4.144 to display the status of LED and rain condition on the serial monitor is shown in Figure 4.146.

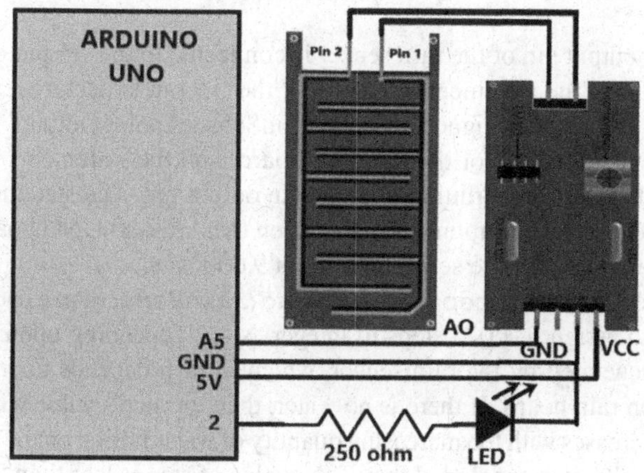

FIGURE 4.144 The interfacing of a rain detector sensor module and LED with Arduino UNO board.

int rainSensorAnalogOutput = A5;	*statement (1)*
int LED=2;	*statement (2)*
void setup()	
{	
pinMode(rainSensorAnalogOutput,INPUT);	*statement (3)*
pinMode(LED,OUTPUT);	*statement (4)*
Serial.begin(9600);	*statement (5)*
}	
void loop()	
{	
int step=analogRead(rainSensorAnalogOutput);	*statement (6)*
if(step< 300)	*statement (7)*
{	
digitalWrite(LED,HIGH);	*statement (8)*
Serial.print("Step of ADC=");	*statement (9)*
Serial.print(step);	*statement (10)*
Serial.print(" LED ON");	*statement (11)*
Serial.println(" RAINING HEAVILY");	*statement (12)*
}	
if (step>= 300)	*statement (13)*
{	
digitalWrite(LED,LOW);	*statement (14)*
Serial.print("Step of ADC=");	*statement (15)*
Serial.print(step);	*statement (16)*
Serial.print(" LED OFF");	*statement (17)*
Serial.println(" MODERATE or NO RAIN");	*statement (18)*
}	
delay(5000);	*statement (19)*
}	

FIGURE 4.145 A program to turn on the LED if it is raining heavily and also display the information on the serial monitor.

Description of the Program:

The analog output pin of the rain sensor is connected to the A5 pin of Arduino board. Using the statements (1) and (3), the A5 pin is declared as an input pin and the name assigned to it is "rainSensorAnalogOutput". The LED is connected to Pin 2 of the Arduino board, and the statements (2) and (4) declare Pin 2 of the Arduino board as an output pin. The statement (5) initializes the serial communication between the Arduino UNO board and the computer to display the serial monitor at 9,600 baud.

The statement (6) `int step=analogRead(rainSensorAnalogOutput)` is used to assign 0–1,023 steps to integer "step" depending upon the analog output generated by the rain sensor, which in turn depends upon the water falling on rain board. If there is no water, then the step's value will be more, and it decreases with the increasing quantity of water falling on the rain board.

Experimentally, it is concluded that if the value of step is less than 300, then it will be heavy rain. Thus, if statement (7) is true, then statements (8) to (12) will be executed, and they will cause LED to turn on, step value, and the

```
COM6 (Arduino/Genuino Uno)

Step of ADC=1023   LED OFF    MODERATE or NO RAIN
Step of ADC=1016   LED OFF    MODERATE or NO RAIN
Step of ADC=993   LED OFF    MODERATE or NO RAIN
Step of ADC=953   LED OFF    MODERATE or NO RAIN
Step of ADC=938   LED OFF    MODERATE or NO RAIN
Step of ADC=938   LED OFF    MODERATE or NO RAIN
Step of ADC=1019   LED OFF    MODERATE or NO RAIN
Step of ADC=1022   LED OFF    MODERATE or NO RAIN
Step of ADC=364   LED OFF    MODERATE or NO RAIN
Step of ADC=353   LED OFF    MODERATE or NO RAIN
Step of ADC=274   LED ON   RAINING HEAVILY
Step of ADC=283   LED ON   RAINING HEAVILY
Step of ADC=283   LED ON   RAINING HEAVILY
Step of ADC=280   LED ON   RAINING HEAVILY
Step of ADC=276   LED ON   RAINING HEAVILY
```

FIGURE 4.146 The snapshot of the serial monitor after the execution of the program as shown in Figure 4.145.

messages "LED ON" and "RAINING HEAVILY" will be displayed on the serial monitor.

If the value of step is greater than or equal to 300, then statement (13) is true, and statements (14) to (18) will be executed. They will cause LED to turn off, step value, and the message "LED OFF" and "MODERATE or NO RAIN" will be displayed on the serial monitor. The statement (19) will generate a delay of 5 seconds.

4.19 INTERFACING AND PROGRAMMING OF ARDUINO UNO WITH ULTRASONIC SENSOR (HC-SR04)

This section shall discuss the interfacing of the ultrasonic sensor with the Arduino UNO board and programming details. The ultrasonic sensor is used to find out how much far an object is from the sensor. The HC-SR04 ultrasonic sensor is used for the description of interfacing circuits and programs. The working and pin description of HC-SR04 ultrasonic sensor is explained in Section 3.18 of Chapter 3.

Program 4.58

Interface an ultrasonic sensor module (HC-SR04) with Arduino UNO board, and write a program to display the distance of an obstacle on the serial monitor.

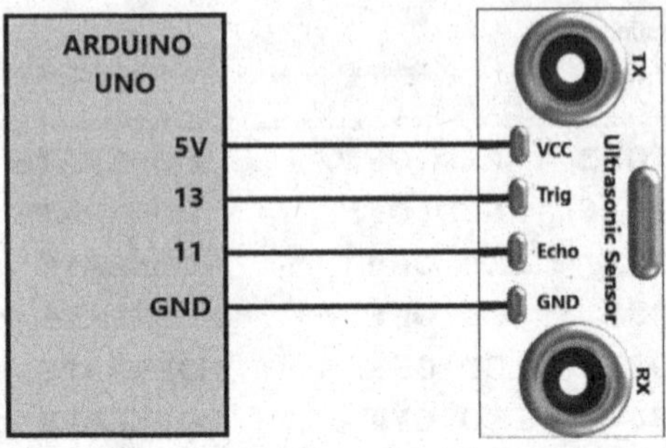

FIGURE 4.147 The interfacing of an ultrasonic sensor module (HC-SR04) with Arduino UNO board.

Solution

The interfacing of the ultrasonic sensor module (HC-SR04) with Arduino UNO board is shown in Figure 4.147. The VCC, Trig, Echo, and GND (ground) pins of ultrasonic sensor module are connected to 5 V, 13, 11, and GND (ground) pins of Arduino board.

An Arduino UNO program to display the distance of an obstacle placed in front of ultrasonic sensor on the serial monitor for the circuit diagram shown in Figure 4.147 is shown in Figure 4.148.

The screenshot of serial monitor of program shown in Figure 4.148 and interfacing circuit shown in Figure 4.147 to display the distance of an obstacle placed in front of ultrasonic sensor on the serial monitor is shown in Figure 4.149.

Description of the Program:

The Trig input pin of the ultrasonic sensor module is connected to Pin 13 of Arduino board. The statement (1) is given name "trigInputPin" to Pin 13 of Arduino board, and the statement (3) declares Pin 13 as an output pin. The Echo output pin of the sensor module is connected to Pin 11 of Arduino board. The statements (2) and (4) give name "echoOutputPin" to Pin 11 of Arduino board and declare it as an input pin, respectively.

The statement (5) initializes the serial communication between Arduino UNO board and the computer to display on the serial monitor at 9,600 baud. The statement (6) declares two float-type variables "timeToAndFroUltrasonicPulse" and "distanceOfObstacle".

In order to initiate the process of measuring the distance of object from the ultrasonic sensor module, a high pulse (5 V) of 10 μs duration is to be applied at the Trig pin of the sensor module. First, by using statements (7) and (8), we shall make sure that "trigInputPin" is getting a low pulse (0 V) for 2 μs. Now by using statements (9) and (10), we shall make "trigInputPin" high for 10 μs. The statement (11) makes "trigInputPin" again low for the next round of calculating the distance of the obstacle. Once the ultrasonic sensor is initiated, the transmitting section of it generates eight pulses.

int trigInputPin= 13;	*statement (1)*
int echoOutputPin= 11;	*statement (2)*
void setup()	
{	
pinMode(trigInputPin,OUTPUT);	*statement (3)*
pinMode(echoOutputPin,INPUT);	*statement (4)*
Serial.begin (9600);	*statement (5)*
}	
void loop()	
{	
float timeToAndFroUltrasonicPulse, distanceOfObstacle;	*statement (6)*
digitalWrite(trigInputPin,LOW);	*statement (7)*
delayMicroseconds(2);	*statement (8)*
digitalWrite(trigInputPin,HIGH);	*statement (9)*
delayMicroseconds(10);	*statement (10)*
digitalWrite(trigInputPin,LOW);	*statement (11)*
timeToAndFroUltrasonicPulse = pulseIn(echoOutputPin,HIGH);	*statement (12)*
*distanceOfObstacle = (timeToAndFroUltrasonicPulse / 2) * 0.0344;*	*statement (13)*
if (distanceOfObstacle >= 400 || distanceOfObstacle <= 2)	*statement (14)*
{	
Serial.print("Distance = ");	*statement (15)*
Serial.print(distanceOfObstacle);	*statement (16)*
Serial.println("Out of range");	*statement (17)*
}	
else	*statement (18)*
{	
Serial.print("Distance = ");	*statement (19)*
Serial.print(distanceOfObstacle);	*statement (20)*
Serial.println(" cm");	*statement (21)*
}	
delay(500);	*statement (22)*
}	

FIGURE 4.148 An Arduino UNO program to display the distance of an obstacle placed in front of ultrasonic sensor on the serial monitor for the circuit diagram shown in Figure 4.147.

```
COM6 (Arduino/Genuino Uno)

Distance = 156.18 cm
Distance = 155.18 cm
Distance = 155.80 cm
Distance = 155.20 cm
Distance = 235.92 cm
Distance = 158.53 cm
Distance = 235.54 cm
Distance = 236.59 cm
Distance = 3.72 cm
Distance = 5.90 cm
Distance = 8.41 cm
Distance = 8.02 cm
Distance = 236.35 cm
Distance = 236.05 cm
Distance = 236.40 cm
```

FIGURE 4.149 The screenshot of serial monitor of program shown in Figure 4.148 and interfacing circuit shown in Figure 4.147 to display the distance of an obstacle placed in front of ultrasonic sensor on the serial monitor.

If there is an obstacle in front of the ultrasonic sensor, then the transmitting section's pulses will get reflected back and received by the receiving section. The Echo pin of the ultrasonic sensor will remain high for the time taken by the eight pulses to travel from the transmitting section and back to the receiving section after getting reflected from the obstacle.

The `pulseIn()` is a function used in statement (12). This function has two parameters: the first parameter is the name of the echo pin, and for the second parameter, we can write either HIGH or LOW. If we write high for the second parameter of `pulseIn()` function, then the pulseIn() function will return the length of time in microseconds for which the Echo pin of sensor is HIGH.[4] The statement (12) will assign the duration of Echo pin for which it is high to "timeToAndFroUltrasonicPulse" variable.

The statement (13) will calculate the distance of the obstacle from the ultrasonic sensor in centimeters. If the distance of the obstacle is greater than or equal to 400 cm or less than or equal to 2 cm, then statement (14) will be valid and statements (15) to (17) will be executed to display the distance of obstacle and "Out of range" message on the serial monitor.

Suppose the distance of the obstacle is less than 400 cm or greater than 2 cm. In that case, statement (18) will be valid, and statements (19) to (21) will be executed to display the distance of obstacle message on the serial monitor. The statement (22) will generate a delay of 500 ms.

4.20 INTERFACING AND PROGRAMMING OF ARDUINO UNO WITH BLUETOOTH MODULE (HC-05)

This section shall discuss the interfacing of Bluetooth module with Arduino UNO board and programming details. The HC-05 Bluetooth module is used for the description of interfacing circuits and programs. The working and pin description of HC-05 Bluetooth module is explained in Section 3.19 of Chapter 3.

Program 4.59

Interface a Bluetooth module (HC-05) with Arduino UNO board, and write a program to display the serial monitor's content when "1" and "2" are pressed in mobile using Bluetooth app.

Solution

The interfacing of the Bluetooth module (HC-05) with Arduino UNO board is shown in Figure 150. The Vcc, GND (ground), TXD, and RXD pins of Bluetooth module are connected to 5 V, GND (ground), 0, and 1 pins of Arduino board.

The TXD and RXD pins of Bluetooth module (HC-05) are connected to the RX (Pin 0) and TX (Pin 1) of Arduino board, respectively.

An Arduino UNO program to display the content on the serial monitor when "1" and "2" are pressed in mobile using Bluetooth app for the circuit diagram shown in Figure 4.150 is shown in Figure 4.151. The screenshot of serial monitor of program shown in Figure 4.151 and interfacing circuit shown in Figure 4.150 to

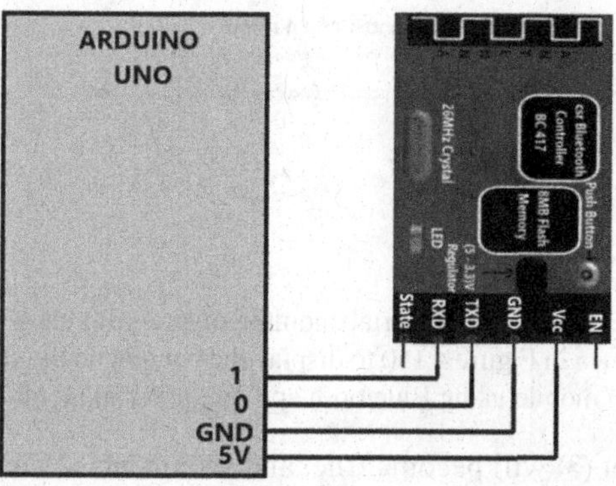

FIGURE 4.150 The interfacing of a Bluetooth module (HC-05) with Arduino UNO board.

int value;	statement (1)
void setup()	
{	
Serial.begin(9600);	statement (2)
}	
void loop()	
{	
if(Serial.available())	statement (3)
{	
value = Serial.read();	statement (4)
Serial.println(value);	statement (5)
}	
}	

FIGURE 4.151 An Arduino UNO program to display the content on the serial monitor when "1" and "2" are pressed in mobile using Bluetooth app for the circuit diagram shown in Figure 4.150.

display the content on the serial monitor when "1" and "2" are pressed in mobile using Bluetooth app on the serial monitor is shown in Figure 4.152.

Description of the Program:

The statement (1) declares an integer-type variable "value". The statement (2) initializes the serial communication between the Arduino UNO board and the computer to display the serial monitor at 9,600 baud.

The `Serial.available()` function returns the number of bytes available in serial port buffer for the read. In serial receiver buffer of Arduino, 64 bytes can be stored. The statement (3) `if(Serial.available())` will return a true value if any data is available in the buffer of receiver port of Arduino; otherwise, it will return false. Therefore, when the Bluetooth module is connected to the Arduino UNO, and "1" or "2" is pressed in mobile using Bluetooth app, the equivalent ASCII code of "1" or "2", which is "49" and "50" will be stored in the buffer of receiver port of Arduino,

```
COM6 (Arduino/Genuino Uno)

49
50
49
50
```

FIGURE 4.152 The screenshot of serial monitor of program shown in Figure 4.151 and interfacing circuit shown in Figure 4.150 to display the content on the serial monitor when "1" and "2" are pressed in mobile using Bluetooth app on the serial monitor.

then statement (3) will become true, and statements (4) and (5) will be executed. The `Serial.read()` function reads the byte available in the buffer of the receiver port of Arduino. The statement (4) will assign the read byte from the buffer of receiver port, i.e., either 49 or 50 to the "value" variable. The statement (5) displays the equivalent ASCII value of pressed key on the serial monitor.

Program 4.60

Interface a Bluetooth module (HC-05) and a LED with Arduino UNO board, and write a program to turn on and off the LED when "1" and "2" are pressed, respectively, in mobile using Bluetooth app.

Solution

The interfacing of the Bluetooth module (HC-05) and LED with Arduino UNO board is shown in Figure 4.153. The VCC, GND (ground), TXD, and RXD pins of Bluetooth module are connected to 5 V, GND (ground), 0, and 1 pins of Arduino board. The anode of LED is connected to Pin 6 of Arduino UNO board through a 250 Ω resistor, and the cathode is connected to the GND (ground) pin of Arduino board.

An Arduino UNO program to turn on and off the LED when "1" and "2" are pressed, respectively, in mobile using Bluetooth app and display the status of LED on serial monitor for the circuit diagram shown in Figure 4.153 is shown in Figure 4.154.

The screenshot of serial monitor of program shown in Figure 4.154 and interfacing circuit shown in Figure 4.153 to display the status of LED when "1" and "2" are pressed, respectively, in mobile using Bluetooth app on the serial monitor is shown in Figure 4.155.

Description of the Program:

The statement (1) declares an integer-type variable "value". The LED is connected to Pin 6 of the Arduino board. The statement (2) assigns the name "LED" to Pin 6 of the Arduino board. The statement (3) declares Pin 6 of the Arduino board as an output pin. The statement (4) initializes the serial communication between the Arduino UNO board and the computer to display the serial monitor at 9,600 baud.

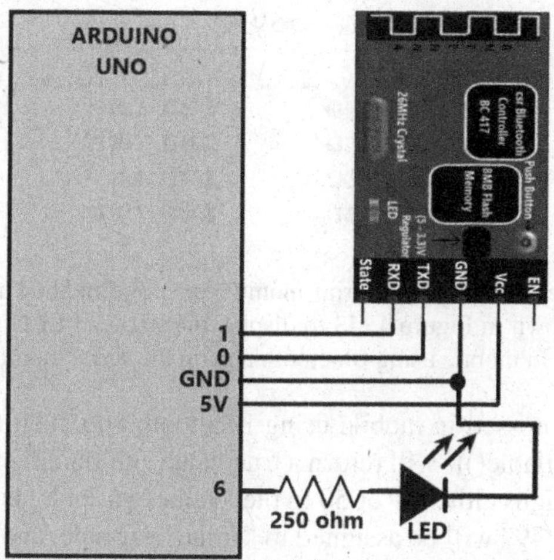

FIGURE 4.153 The interfacing of a Bluetooth module (HC-05) and a LED with Arduino UNO board.

int value;	*statement (1)*
int LED=6;	*statement (2)*
void setup()	
{	
pinMode(LED,OUTPUT);	*statement (3)*
Serial.begin(9600);	*statement (4)*
}	
void loop()	
{	
if(Serial.available())	*statement (5)*
{	
value = Serial.read();	*statement (6)*
if (value==49)	*statement (7)*
{	
digitalWrite(LED,HIGH);	*statement (8)*
Serial.print("1 IS PRESSED");	*statement (9)*
Serial.println(" LED ON");	*statement (10)*
}	
else if (value==50)	*statement (11)*
{	
digitalWrite(LED,LOW);	*statement (12)*
Serial.print("2 IS PRESSED");	*statement (13)*
Serial.println(" LED OFF");	*statement (14)*
}	
}	
}	

FIGURE 4.154 An Arduino UNO program to turn on and off the LED when "1" and "2" are pressed, respectively, in mobile using Bluetooth app and display the status of LED on serial monitor for the circuit diagram shown in Figure 4.153.

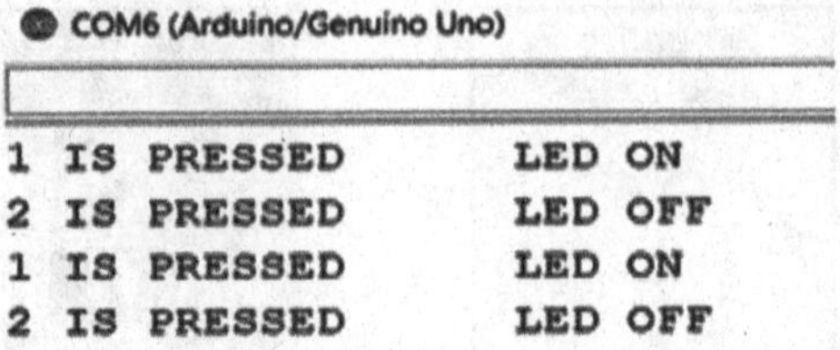

```
COM6 (Arduino/Genuino Uno)

1 IS PRESSED          LED ON
2 IS PRESSED          LED OFF
1 IS PRESSED          LED ON
2 IS PRESSED          LED OFF
```

FIGURE 4.155 The screenshot of serial monitor of program shown in Figure 4.154 and interfacing circuit shown in Figure 4.153 to display the status of LED when "1" and "2" are pressed, respectively, in mobile using Bluetooth app on the serial monitor.

If "1" or "2" is pressed in mobile using Bluetooth app, then the statement (5) "if(Serial.available())" will return a true value and statement (6) will be executed and assigns either 49 or 50 to the "value" variable. If "1" is pressed in mobile, then "49" will be assigned to "value" variable, and if "2" is pressed in mobile, then "50" will be assigned to "value" variable.

If key "1" was pressed, then statement (7) will become true and statements (8), (9), and (10) will be executed causing LED to turn on and messages "1 is PRESSED" and "LED ON" will be displayed on the serial monitor.

If key "2" was pressed, then statement (11) will become true and statements (12), (13), and (14) will be executed causing LED to turn off and messages "2 is PRESSED" and "LED OFF" will be displayed on the serial monitor.

4.21 INTERFACING AND PROGRAMMING OF ARDUINO UNO WITH GSM MODULE (SIM900A)

This section shall discuss the interfacing of GSM module with Arduino UNO board and programming details. The GSM module is used for communication with remote locations provided the mobile tower is in range. The SIM900A GSM module is used for the description of interfacing circuits and programs. The working and pin description of SIM900A GSM module is explained in Section 3.20 of Chapter 3.

Program 4.61

Interface a SIM900A GSM module and LED with Arduino UNO board, and write a program to turn on and off the LED when we send SMS "1" and "2", respectively, and display the LED position serial monitor.

Solution

The interfacing of the SIM900A GSM module and LED with Arduino UNO board is shown in Figure 4.156. The GND (ground), TXD, and RXD pins of GSM module are connected to GND (ground), 0, and 1 pins of Arduino board. The anode of LED is connected to the pin number 13 of Arduino UNO board through a 250 Ω resistor, and the cathode is connected to the GND (ground) pin of Arduino board. A 5 V, 2 A adapter powers the GSM module. The TXD and RXD pins of the GSM module are connected to the RX (Pin 0) and TX (Pin 1) of Arduino board.

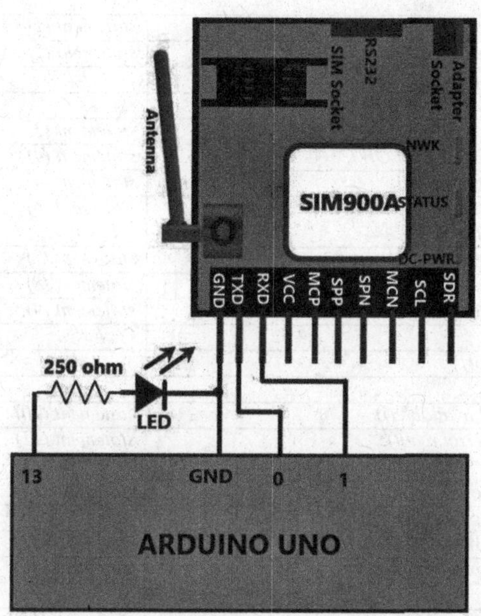

FIGURE 4.156 The interfacing of a SIM900A GSM module with Arduino UNO board.

An Arduino UNO program to turn on and off the LED when we send SMS "1" and "2", respectively, and display the status of LED on the serial monitor for the circuit diagram shown in Figure 4.156 is shown in Figure 4.157.

The screenshot of serial monitor of program shown in Figure 4.157 and interfacing circuit shown in Figure 4.156 to display the status of LED when we send SMS "1" and "2" on the serial monitor is shown in Figure 4.158.

Description of the Program:

The LED is connected to Pin 13 of the Arduino board. The statement (2) assigns the name "LED" to Pin 13 of the Arduino board. The statement (3) declares Pin 13 of the Arduino board as an output pin. The statement (4) initially turns off the LED. The statement (5) initializes the serial communication between the Arduino UNO board and the computer to display the serial monitor at 9,600 baud.

The statement (6) *Serial.println("AT+CMGF=1")* will set the GSM module in SMS text mode. *Serial.println("AT+CNMI=2,2,0,0,0")* will enable GSM module to receive SMS. The statement (10) *if(Serial.available())* will return a true value if any data is available in the buffer of receiver port of Arduino; otherwise, it will return false. Therefore, when the GSM module is received an SMS "1" or "2", the equivalent ASCII code of "1" or "2" will be stored in the buffer of receiver port of Arduino and statement (10) will become true and statement (11) *value=Serial.read()* will be executed. The *Serial.read()* function reads the byte available in the buffer of the receiver port of Arduino and assigns it to the

char value;	*statement (1)*
int LED=13;	*statement (2)*
void setup()	
{	
pinMode(LED,OUTPUT);	*statement (3)*
digitalWrite(LED,LOW);	*statement (4)*
Serial.begin(9600);	*statement (5)*
Serial.println("AT+CMGF=1");//set sms mode to text	*statement (6)*
delay(100);	*statement (7)*
Serial.println("AT+CNMI=2,2,0,0,0");	*statement (8)*
delay(100);	*statement (9)*
}	
void loop()	
{	
if (Serial.available())	*statement (10)*
value=Serial.read();	*statement (11)*
{	
if (value=='1')	*statement (12)*
{	
digitalWrite(LED,HIGH);	*statement (13)*
Serial.println("LED ON");	*statement (14)*
}	
else if(value=='2')	*statement (15)*
{	
digitalWrite(LED,LOW);	*statement (16)*
Serial.println("LED OFF");	*statement (17)*
}	
delay(5000);	*statement (18)*
}	
}	

FIGURE 4.157 An Arduino UNO program to turn on and off the LED when we send SMS "1" and "2", respectively, and display the status of LED on the serial monitor for the circuit diagram shown in Figure 4.156.

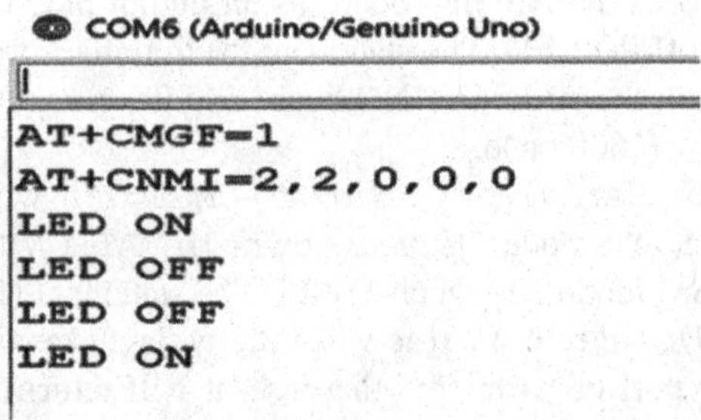

FIGURE 4.158 The screenshot of serial monitor of program shown in Figure 4.157 and interfacing circuit shown in Figure 4.156 to display the status of LED when we send SMS "1" and "2" on the serial monitor.

"value" variable. If the SMS received is "1", then the statement (12) will become true, and statements (13) and (14) will be executed. The statement (13) and (14) will turn on the LED connected to Pin 13 of the Arduino board, and "LED ON" message will be displayed on the serial monitor.

If the SMS received is "2", then the statement (15) will become true instead of (12). If statement (15) is true, then the statements (16) and (17) will be executed. The statements (16) and (17) will turn off the LED connected to Pin 13 of the Arduino board, and "LED OFF" message will be displayed on the serial monitor.

The statement (18) is used to generate a delay of 5 seconds.

4.22 INTERFACING AND PROGRAMMING OF ARDUINO UNO USING I2C PROTOCOL

In this section, we shall discuss the interfacing of two Arduino UNO boards using Inter-Integrated Circuit (I2C) Communication Protocol. The I2C protocol is used for serial data transfer among more than two devices. The concept of I2C protocol is explained in Section 1.7 of Chapter 1.

Program 4.62

Interface two Arduino Boards using I2C Protocol. A LED and a switch connected to the Arduinos using I2C protocols are shown in Figure 4.159. Write a program to turn on the LED whenever the switch is pressed.

Solution

The interfacing of switch, LED, and two Arduino UNO boards as master and slave using I2C Protocol is shown in Figure 4.159. The A4 and A5 pins of master and slave Arduino boards are connected with each other. The anode of LED is connected to Pin 7 of the slave Arduino UNO board through a 250 Ω resistor, and the cathode is connected to the GND (ground) pin of slave Arduino board. The Terminal T1 of the push-button switch is connected to the GND (ground) pin of master Arduino board, and the Terminal T2 is connected to the one terminal of 1 KΩ resistor. The other terminal of 1 KΩ resistor is connected to the 5 V pin of master Arduino board. The junction of terminal T2 of switch and 1 KΩ resistor is extended and connected to the Pin 6 of master Arduino board.

During programming, the Arduino in which LED is connected will be declared as slave and the Arduino in which switch is connected will be declared as master. The two Arduinos are connected by using I2C protocol, and whenever switch connected to master Arduino is pressed, the LED connected to the slave Arduino will turn on.

A master Arduino and a slave Arduino program to turn on the LED connected to slave Arduino whenever switch connected to master Arduino is pressed are shown in Figures 4.160 and 4.161, respectively.

Description of the program for master Arduino is shown in Figure 4.160:

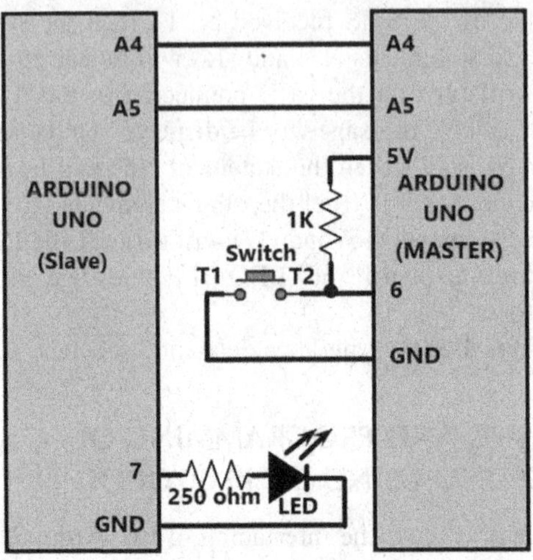

FIGURE 4.159 The interfacing of two Arduino boards, LED and switch using I2C Protocol.

#include < Wire.h >	*statement (1)*
int x =0;	*statement (2)*
int slaveAddress = 9;	*statement (3)*
int pushButton = 6;	*statement (4)*
int buttonState = 0;	*statement (5)*
void setup()	
{	
pinMode(pushButton,INPUT);	*statement (6)*
Serial.begin(9600);	*statement (7)*
Wire.begin();	*statement (8)*
}	
void loop()	
{	
buttonState = digitalRead(pushButton);	*statement (9)*
if (buttonState == 1)	*statement (10)*
{	
x = 1;	*statement (11)*
}	
else if (buttonState == 0)	*statement (12)*
{	
x = 0;	*statement (13)*
}	
Wire.beginTransmission(slaveAddress);	*statement (14)*
Wire.write(x);	*statement (15)*
Wire.endTransmission();	*statement (16)*
delay(200);	*statement (17)*
}	

FIGURE 4.160 An Arduino UNO program for master Arduino to turn on the LED connected to slave Arduino whenever switch connected to master Arduino is pressed for the circuit diagram shown in Figure 4.159.

#include < Wire.h >	statement (1)
int LED = 13;	statement (2)
int x = 0;	statement (3)
void setup()	
{	
pinMode (LED,OUTPUT);	statement (4)
Wire.begin(9);	statement (5)
Wire.onReceive(receiveEvent);	statement (6)
}	
void receiveEvent(int bytes)	statement (7)
{	
x = Wire.read();	statement (8)
Wire.endTransmission();	statement (9)
}	
void loop()	
{	
if (x == 0)	statement (10)
{	
digitalWrite(LED, HIGH);	statement (11)
delay(200);	statement (12)
}	
else if (x == 1)	statement (13)
{	
digitalWrite(LED, LOW);	statement (14)
delay(200);	statement (15)
}	
}	

FIGURE 4.161 An Arduino UNO program for slave Arduino to turn on the LED connected to slave Arduino whenever switch connected to master Arduino is pressed for the circuit diagram shown in Figure 4.159.

The statement (1) `#include < Wire.h >` is used to include wire library. The statement (2) declares an integer-type variable "x" with its initial value 0. The statement (3) `int slaveAddress = 9;` declares an integer-type variable `slaveAddress` with its initial value 9. We will assign the address to slave Arduino as 9 in later part of program. Using the statements (4) and (6), Pin 6 of Arduino is given a name `pushButton` and declared an input pin. The statement (5) `int buttonState = 0;` declares an integer-type variable `buttonState` with its initial value 0. The `Serial.begin(9600)` function of the statement (7) will initialize the serial communication at 9,600 Baud. The statement (8) `Wire.begin()` initiates the I2C communication at Pins A4 and A5 of Arduino board. Since in `Wire.begin()` function no address is mentioned with in the brackets, the Arduino will join as a master. The statement (9) `buttonState = digitalRead(pushButton);` will read the digital value of "pushButton" (Pin 6) and assign this value to "buttonState". As per Figure 4.159 if the push button is not pressed, then "1" (5 V) will be assigned to variable "buttonState" and if the push button is pressed, then "0" (0 V) will be assigned to variable "buttonState".

If the switch is not pressed, then statement (10) `if (buttonState == 1)` is true and statement (11) will update variable x to 1. If the switch is pressed, then statement (12) `else if (buttonState == 0)` is true and statement (13) will update variable x to 0. The statement (14) `Wire.beginTransmission(slaveAddress);` begins the transmission with slaveArduino whose address is defined as 9. The statement (15) `Wire.write(x);` queues the value of x for transmission. The transmission of bytes initiated using `Wire.beginTransmission(slaveAddress)` function is ended using statement (16) `Wire.endTransmission().` The `delay(200)` function in statement (17) generates a delay of 200 ms.

Description of the program for slave Arduino is shown in Figure 4.161:

The statement (1) `#include < Wire.h >` is used to include wire library. Using the statements (2) and (4), Pin 13 of Arduino is given a name "LED" and declared an output pin. The statement (3) declares an integer-type variable "x" with its initial value 0. The statement (5) `Wire.begin(9)` initiates the wire library and Arduino joins the I2C bus as a slave with address 9.

The reciveEvent () function is defined to read the value 0 or 1, which is sent by Master Arduino and stores this value in variable x by using statements (6) to (8). The statement (9) `Wire.endTransmission()` stops the transmission.

Due to statements (10) and (11), the LED connected to Pin 13 of slave Arduino will turn on if the switch is pressed. Due to statements (13) and (14), the LED will turn off if the switch is not pressed. The `delay(200)` function in statement (12) and statement (15) generate a delay of 200 ms.

5 Arduino-Based Projects

5.1 ARDUINO-BASED OBSTACLE DETECTION AND WARNING SYSTEM

This project aims to generate a warning sound whenever an obstacle is detected within 400 cm range from an ultrasonic sensor module (HC-SR04).

Readers can refresh the concepts by revisiting the detailed explanation of the ultrasonic sensor module (HC-SR04) given in Section 3.18 of Chapter 3 and the program related to the ultrasonic sensor in Section 4.19 of Chapter 4.

The components required for this project are as follows:

i. Arduino UNO Board
ii. Generic breadboard
iii. Ultrasonic sensor (HC-SR04)
iv. 5 V buzzer
v. Connecting wires
vi. Arduino cable
vii. 100 Ω resistor.

Description of the Circuit:

The interfacing of the ultrasonic sensor module (HC-SR04) and a 5V buzzer with Arduino UNO board to detect an obstacle and generate a warning sound whenever an obstacle is detected is shown in Figure 5.1. The ultrasonic sensor module (HC-SR04) is used to detect an obstacle coming in front of it. A 5V buzzer is used to generate a warning sound whenever an obstacle is detected in front of the ultrasonic sensor. The buzzer's positive end (T2) is connected to Pin 5 of Arduino board through a 100 Ω resistor, and the other terminal of the buzzer (T1) is connected to the GND (ground) pin of Arduino board. The VCC and GND pins of the ultrasonic sensor module (HC-SR04) are connected to 5 V and GND pins of Arduino board, respectively. The Trig input pin and the Echo output pin of the HC-SR04 ultrasonic sensor module are connected to Pin 13 and Pin 11 of Arduino board.

The flowchart of working principle of the obstacle detection system with warning sound generation is shown in Figure 5.2.

An Arduino UNO program to detect an obstacle and generate a warning sound whenever an obstacle is detected within 400 cm range in front of an ultrasonic sensor for the circuit diagram shown in Figure 5.1 is shown in Figure 5.3.

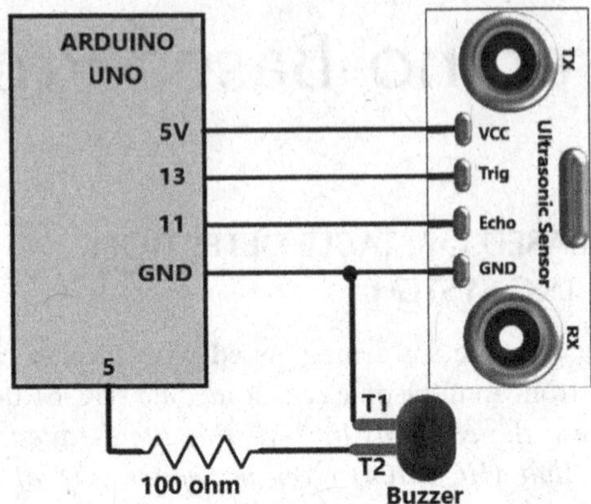

FIGURE 5.1 The circuit diagram to detect an obstacle and generate a warning sound whenever an obstacle is detected.

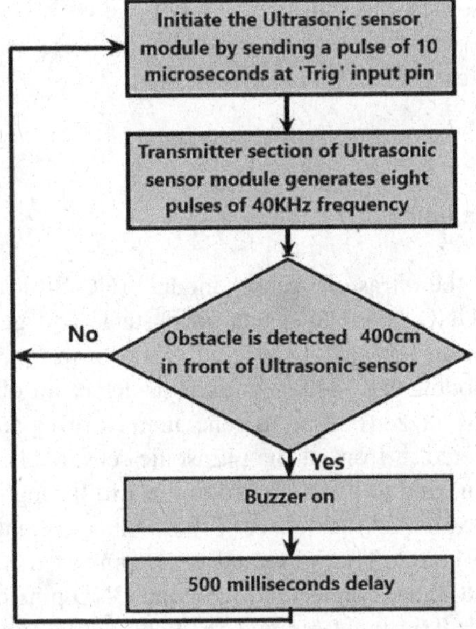

FIGURE 5.2 The flowchart of working principle of the obstacle detection system with warning sound generation.

Description of the Program:

The Trig input pin of the ultrasonic sensor module is connected to Pin 13 of the Arduino board. The statement (1) gives the name "trigInputPin" to Pin 13 of the Arduino board, and the statement (4) declares Pin 13 as an output pin. The Echo output pin of the sensor module is connected to Pin 11 of the

int trigInputPin= 13;	*statement (1)*
int echoOutputPin= 11;	*statement (2)*
int buzzer= 7;	*statement (3)*
void setup()	
{	
pinMode(trigInputPin,OUTPUT);	*statement (4)*
pinMode(echoOutputPin,INPUT);	*statement (5)*
pinMode(buzzer,OUTPUT);	*statement (6)*
}	
void loop()	
{	
float timeToAndFroUltrasonicPulse, distanceOfObstacle;	*statement (7)*
digitalWrite(trigInputPin,LOW);	*statement (8)*
delayMicroseconds(2);	*statement (9)*
digitalWrite(trigInputPin,HIGH);	*statement (10)*
delayMicroseconds(10);	*statement (11)*
digitalWrite(trigInputPin,LOW);	*statement (12)*
timeToAndFroUltrasonicPulse = pulseIn(echoOutputPin,HIGH);	*statement (13)*
*distanceOfObstacle = (timeToAndFroUltrasonicPulse / 2) * 0.0344;*	*statement (14)*
if (distanceOfObstacle <= 400)	*statement (15)*
{	
digitalWrite(buzzer,HIGH);	*statement (16)*
}	
else	*statement (17)*
{	
digitalWrite(buzzer,LOW);	*statement (18)*
}	
delay(500);	*statement (19)*
}	

FIGURE 5.3 An Arduino UNO program to detect an obstacle and generate a warning sound whenever an obstacle is detected for the circuit diagram shown in Figure 5.1.

Arduino board. The statements (2) and (5) give the name "echoOutputPin" to Pin 11 of the Arduino board and declare it an input pin. The statements (3) and (6) give the name "buzzer" to Pin 5 of the Arduino board and declare it an output pin. The statement (7) declares two float-type variables, namely, "timeToAndFroUltrasonicPulse" and "distanceOfObstacle".

To initiate measuring the object's distance from the ultrasonic sensor module, a HIGH pulse (5 V) of 10 μs duration is applied at the Trig pin of the sensor module. First, by using statements (8) and (9), we shall make sure that "trigInputPin" is getting a low pulse (0 V) for 2 μs. Using statements (10) and (11), we shall make "trigInputPin" high (5 V) for 10 μs. The statement (12) makes "trigInputPin" again low for the next round of calculating the obstacle's distance. Once the ultrasonic sensor is initiated, the transmitting section of it generates eight pulses. Suppose there is an obstacle in front of the ultrasonic sensor up to 400 cm. In that case, the transmitting section's pulses will get reflected back and received by the receiving section.

The Echo pin of the ultrasonic sensor will remain high for the time taken by the eight pulses to travel from the transmitting section and back to the receiving section after getting reflected from the obstacle.

The `pulseIn()` is a function used in statement (13). This function has two parameters: the first parameter is the name of the echo pin, and for the second parameter, we can write either HIGH or LOW. If we write HIGH for the second parameter of `pulseIn()` function, then the `pulseIn()` function will return the length of time in microseconds for which the Echo pin of sensor is HIGH.[4] The statement (13) will assign the duration of Echo pin for which it is HIGH to "timeToAndFroUltrasonicPulse" variable. The statement (14) will calculate the distance of the obstacle from the ultrasonic sensor in centimeters. If the distance of the obstacle is less than or equal to 400 cm, then statements (15) and (16) will trigger the buzzer, and a warning sound will be generated. The statement (19) will generate a delay of 500 ms.

5.2 ARDUINO-BASED GAS LEAKAGE DETECTION

This project aims to generate a warning sound whenever gas leakage is detected by the gas sensor (MQ2).

Readers can refresh the concepts by revisiting the detailed explanation of gas sensor (MQ2) given in Section 3.16 of Chapter 3 and program related to gas sensor (MQ2) in Section 4.17 of Chapter 4.

The components required for this project are as follows:

i. Arduino UNO Board
ii. Generic breadboard
iii. Gas sensor (MQ2)
iv. 5 V buzzer
v. Connecting wires
vi. Arduino cable
vii. 100 Ω resistor.

Description of the Circuit:

The circuit diagram to generate a warning sound whenever the gas sensor detects gas leakage is shown in Figure 5.4. The gas sensor module (MQ2) is used to detect the leakage of gas. A 5V buzzer is used to generate a warning sound whenever the leakage of gas is detected. The buzzer's positive end (T2) is connected to Pin 7 of Arduino board through a 100 Ω resistor, and the other terminal of the buzzer (T1) is connected to the GND (ground) pin of Arduino board. The Vcc and GND pins of the gas sensor module (MQ2) are connected to 5 V and GND (ground) pins of Arduino board. The analog output pin of the sensor module is connected to the analog input pin A0 of the Arduino board.

The flowchart of working principle of gas leakage detection system with warning sound generation is shown in Figure 5.5.

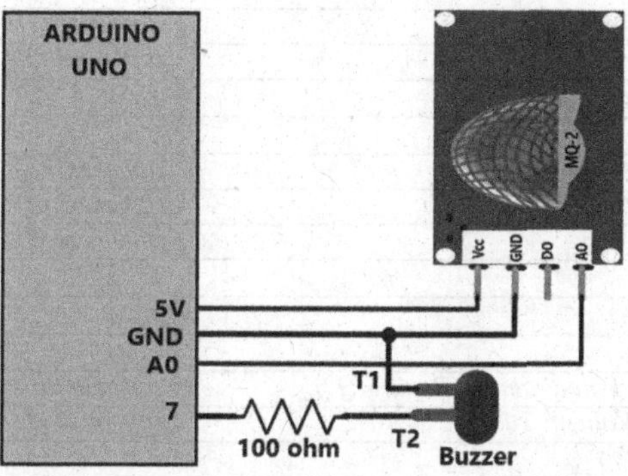

FIGURE 5.4 The circuit diagram to detect gas leakage and generate a warning sound whenever gas leakage is detected.

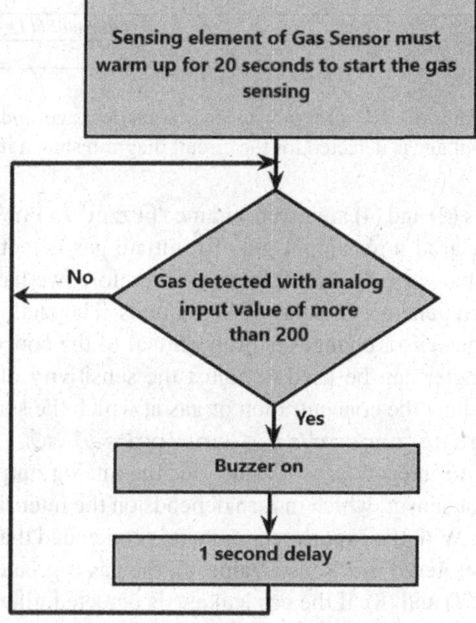

FIGURE 5.5 The flowchart of working principle of gas leakage detection system with warning sound generation.

An Arduino UNO program to generate a warning sound whenever gas leakage is detected from gas leakage sensor module is shown in Figure 5.6.

Description of the Program:

Since the analog output pin of smoke detector sensor is connected to analog pin A0 Arduino board and the sensor's output voltage level will enter inside Arduino from the A0 pin, it has to be an input pin. Using the statements (1) and (3), the A0 pin is given a name "MQ2" and declared as an input pin.

int MQ2=A0;	*statement (1)*
int buzzer=7;	*statement (2)*
void setup()	
{	
pinMode(MQ2,INPUT);	*statement (3)*
pinMode(buzzer,OUTPUT);	*statement (4)*
delay(20000);	*statement (5)*
}	
void loop()	
{	
int sensorValue= analogRead(MQ2);	*statement (6)*
if(sensorValue > 200)	*statement (7)*
{	
digitalWrite(buzzer,HIGH);	*statement (8)*
}	
else	*statement (9)*
{	
digitalWrite(buzzer,LOW);	*statement (10)*
}	
delay(1000);	*statement (11)*
}	

FIGURE 5.6 An Arduino UNO program to detect gas leakage and generate a warning sound whenever gas leakage is detected for the circuit diagram shown in Figure 5.4.

The statements (2) and (4) are given a name "buzzer" to Pin 7 of the Arduino board and declared it an output pin. To initiate gas detection, the sensing element must be warm-up for 20 seconds; therefore, we have written statement (5), which generates a delay of 20 seconds. The analog output voltage provided by the sensor changes is proportional to the concentration of gas. The potentiometer can be used to adjust the sensitivity of the sensor. We can use it to adjust the concentration of gas at which the sensor detects it.[3]

The statement (6) `int sensorValue=analogRead(MQ2)` is used to assign 0–1,023 steps to integer "sensorValue" for the analog output generated by the gas detector sensor, which in turn depends on the intensity of gas around the gas sensor. With the experimentation, it is concluded that sound warning should be generated if the "sensorValue" of the gas is greater than 200. Due to statements (7) and (8), if the gas leakage is detected, then the buzzer will turn on; otherwise, due to statements (9) and (10), it will turn off. The statement (11) will generate a delay of 1 second.

5.3 ARDUINO-BASED BURGLAR DETECTION

This project aims to send an SMS whenever someone touches the touch sensor.

Readers can refresh the concepts by re-visiting the detailed explanation of touch sensor and GSM module given in Sections 3.15 and 3.20 of Chapter 3, respectively, and program related to touch sensor and GSM module in Sections 4.16 and 4.21 of Chapter 4, respectively.

The components required for this project are as follows:

i. Arduino UNO Board
ii. Generic breadboard
iii. Capacitive touch sensor
iv. GSM module
v. Connecting wires
vi. Arduino cable.

Description of the Circuit:

The interfacing of a capacitive touch sensor module and SIM900A GSM module with Arduino UNO board is shown in Figure 5.7. The VCC and GND (ground) pins of touch sensor are connected to 5 V and GND (ground) pins of Arduino UNO board. The digital output pin of the touch sensor (SIG) is connected to Pin 11 of the Arduino board. A 5V, 2A adapter powers the GSM module. The GND (ground) pin of the GSM module is connected to the GND (ground) pin of the Arduino board. The TXD and RXD pins of the GSM module are connected to the RX (Pin 0) and TX (Pin 1) of Arduino board respectively.

The flowchart and the project's program to send an SMS whenever someone touches the touch sensor are shown in Figures 5.8 and 5.9.

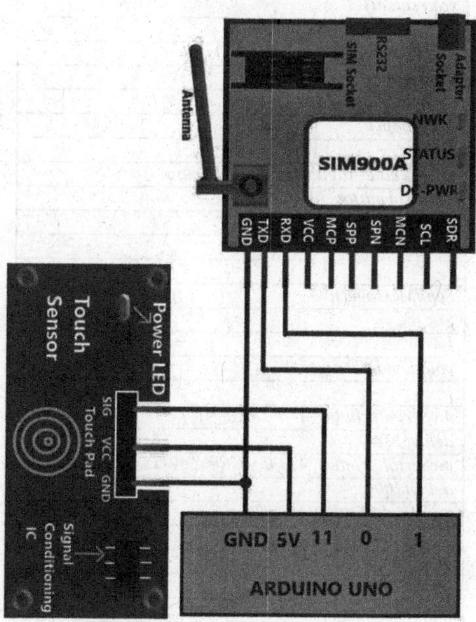

FIGURE 5.7 The circuit diagram to send an SMS whenever someone touches the touch sensor.

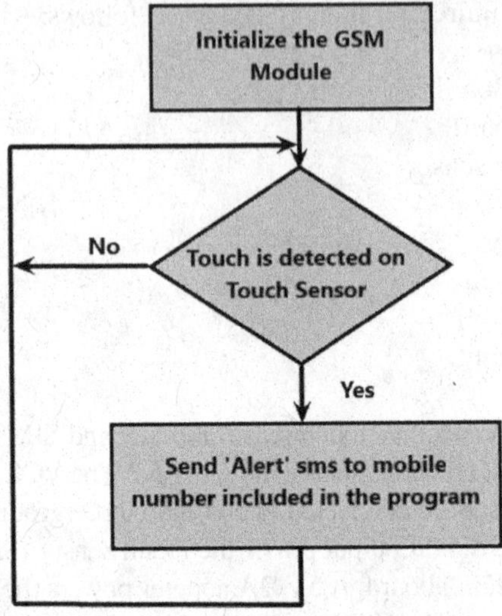

FIGURE 5.8 The flowchart of working principle of the project to send an SMS whenever someone touches the touch sensor.

```
#include <SoftwareSerial.h>
SoftwareSerial mySerial(10,11);
int touchValue=0;
int touchSensor= 11;
void setup()
{
pinMode(touchSensor,INPUT);
mySerial.begin(9600);
}
void loop()
{
touchValue=digitalRead(touchSensor);
Serial.print(touchValue);
Serial.print("\n");
if (touchValue==1)
{
SendMessage();
}
}
void SendMessage()
{
mySerial.println("AT+CMGF=1");
delay(1000);
mySerial.println("AT+CMGS=\"+91xxxxxxxxxx\"\r");
delay(1000);
mySerial.println("ALERT");
delay(100);
mySerial.println((char)9);
delay(1000);
}
```

FIGURE 5.9 An Arduino UNO program to send an SMS whenever someone touches the touch sensor for the circuit diagram shown in Figure 5.7.

Description of the Program:

The program is developed so that if there is a touch on the capacitive touch sensor, then the GSM module will send an alert message to the mobile number included in the program.

The digital output pin of the capacitive touch sensor is connected to Pin 11 of the Arduino board. The Pin 11 of the Arduino board is declared as an input pin, and the name assigned to Pin 11 is "touchsensor". The output voltage level of the sensor is assigned to an integer variable "touchValue". If we touch the sensor's touchpad, then the digital output pin of the capacitive touch sensor module generates 5 V (Logic 1); otherwise, it generates 0 V (Logic 0), and this value is stored in variable "touchValue". If the sensor is touched, then sensor's output pin generates 5 V (Logic 1), and Pin 11 of Arduino receives Logic 1, and it will cause the GSM module to send the "ALERT" message to the mobile number included in the program.

5.4 ARDUINO-BASED WEATHER MONITORING SYSTEM

This project aims to display the information of temperature, humidity, and rain on LCD.

Readers can refresh the concepts by revisiting the detailed explanation of LCD module, potentiometer, humidity and temperature sensor (DHT11), and rain detector given in Sections 3.4, 3.5, 3.9, and 3.17 of Chapter 3, respectively, and program related to LCD module, potentiometer, humidity and temperature sensor (DHT11), and rain detector in Sections 4.6, 4.10, and 4.18 of Chapter 4, respectively.

The components required for this project are as follows:

i. Arduino UNO Board
ii. Generic breadboard
iii. LCD module
iv. Humidity and temperature sensor (DHT11)
v. Rain detector (FC-07)
vi. 5/10 KΩ potentiometer
vii. Connecting wires
viii. Arduino cable.

Description of the Circuit:

The interfacing of the rain detector sensor module (FC-07), DHT11 temperature and humidity sensor, and LCD with Arduino UNO board is shown in Figure 5.10. The pin numbers 1, 2, and 4 of DHT11 are connected to the 5 V, A0, and GND (ground) pin of Arduino board. The Vcc, GND (ground), and AO pins of rain detector sensor module are connected to the 5 V, GND (ground), and A5 pins of Arduino board. The VDD and +5 V pin of LCD is connected to the 5 V pin of Arduino board. The VSS, GND, and RW

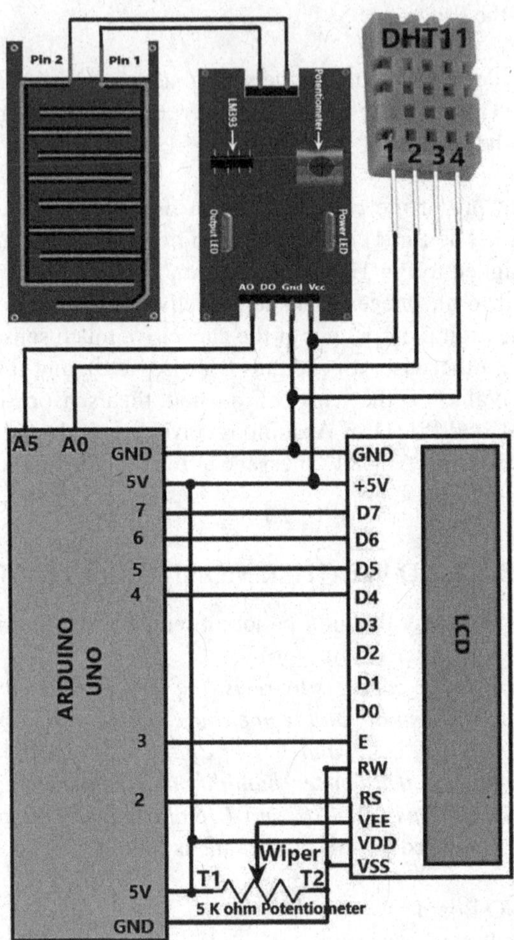

FIGURE 5.10 The circuit diagram to display the information of temperature, humidity, and rain on LCD.

(Read/Write) pin of LCD is connected to the GND (ground) pin of Arduino board. The RS (Register Select) pin of LCD is connected to the pin number 2 of Arduino board. The EN (Enable) pin of LCD is connected to the pin number 3 of Arduino board. The terminals T1 and T2 of 5 KΩ potentiometer are connected to the 5 V and GND (ground) pin of Arduino board. The VEE pin of LCD is connected to the middle (wiper) terminal of 5 KΩ potentiometer. The D4–D7 pins of LCD are connected to the pin numbers 4–7 of Arduino board.

The pin-to-pin mapping of an LCD and Arduino UNO board is shown in Table 4.4 in Section 4.6 of Chapter 4.

The flowchart and the project's program to display the information of temperature, humidity, and rain on LCD are shown in Figures 5.11 and 5.12, respectively.

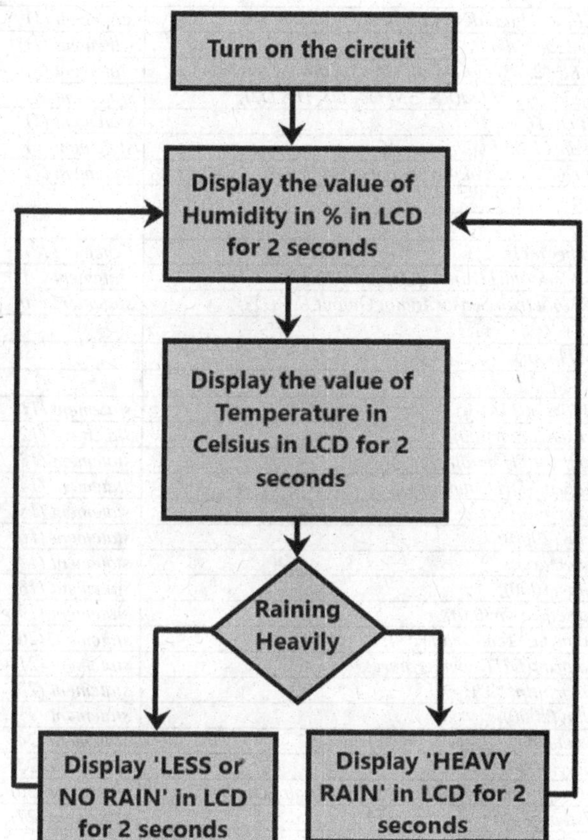

FIGURE 5.11 The flowchart of working principle of the project to display the information of temperature, humidity, and rain on LCD.

Description of the Program:

The statement (1) `#include <LiquidCrystal.h>` is used to include LCD library. The statement (3) `int RS=2, EN=3, D4=4, D5=5, D6=6, D7=7` initializes LCD pins with Arduino UNO pins as shown in Figure 5.10. The statement (4) `LiquidCrystal lcd(RS, EN, D4, D5, D6, D7)` creates an object "lcd" with pin names (RS, EN, D4, D5, D6, D7). The statement (3) has already assigned the pin names of "lcd" object, as shown in Table 4.5. The statement (8) `lcd.begin(16, 2)` is used to initialize the "lcd" object created in the statement (4) as a 16 column 2 row LCD. The statement "lcd.setCursor(3,0)" appeared at statements (12), (19), (28), and (31) is used to initialize displaying information of humidity value, temperature value, and rain condition from 0th column and 0th row on LCD module, respectively.

The statement (2) `#include <dht.h>` includes the dht library. The dht library has all the functions required to get the humidity and temperature

include <LiquidCrystal.h>	statement (1)
#include <dht.h>	statement (2)
int RS=2, EN=3, D4=4, D5=5, D6=6, D7=7;	statement (3)
LiquidCrystal lcd(RS, EN, D4, D5, D6, D7);	statement (4)
dht DHT;	statement (5)
int dht11AnalogOnput =A0;	statement (6)
int rainSensorAnalogOutput = A5;	statement (7)
void setup()	
{	
lcd.begin(16, 2);	statement (8)
pinMode(dht11AnalogOnput,INPUT);	statement (9)
pinMode(rainSensorAnalogOutput,INPUT);	statement (10)
}	
void loop()	
{	
DHT.read11(A0);	statement (11)
lcd.setCursor(0,0);	statement (12)
lcd.print("Humidity ");	statement (13)
lcd.print(DHT.humidity);	statement (14)
lcd.println(" %");	statement (15)
delay(2000);	statement (16)
lcd.clear();	statement (17)
delay(1000);	statement (18)
lcd.setCursor(0,0);	statement (19)
lcd.print("Temperature ");	statement (20)
lcd.print(DHT. temperature);	statement (21)
lcd.println("C");	statement (22)
delay(2000);	statement (23)
lcd.clear();	statement (24)
delay(1000);	statement (25)
int step=analogRead(rainSensorAnalogOutput);	statement (26)
if(step< 300)	statement (27)
{	
lcd.setCursor(0,0);	statement (28)
lcd.print("HEAVY RAIN");	statement (29)
}	
if (step>= 300)	statement (30)
{	
lcd.setCursor(0,0);	statement (31)
lcd.print("LESS or NO RAIN");	statement (32)
}	
lcd.clear();	statement (33)
delay(2000);	
}	

FIGURE 5.12 An Arduino UNO program to display the information of temperature, humidity, and rain on LCD for the circuit diagram as shown in Figure 5.10.

readings from the sensor. The statement (5) `dht DHT` creates an object of name DHT.

The statements (6) and (9) initialize the analog input pin A0 as input and assigned a name `dht11AnalogOnput` to it. The statements (7) and (10) initialize the analog input pin A5 as input and assigned a name “rain sensor-AnalogOutput” to it.

The statement (11) `DHT.read11(A0)` reads the value of humidity and temperature from analog pin A0 and assigns its value to object DHT. The humidity value can be accessed by `DHT.humidity` function, and the temperature value can be accessed by `DHT.temperature` function.

The statements from (12) to (16) display the humidity value in % in the LCD module for 2 seconds. The statements (17) and (18) will clear the LCD module.

The statements from (19) to (23) are used to display Celsius's temperature value in the LCD module for 2 seconds. The statements (24) and (25) will clear the LCD module.

The analog output pin of the rain sensor is connected to the A5 pin of Arduino board. Using the statements (7) and (10), the A5 pin is declared as an input pin and the name assigned to it is "rainSensorAnalogOutput".

The statement (26) `int step=analogRead(rainSensorAnalogOutput)` is used to assign (0–1,023) steps to integer "step" depending upon the analog output generated by the rain sensor, which in turn depends upon the water falling on rain board. If there is no water, then the step's value will be more, and it decreases with the increasing quantity of water falling on the rain board.

Experimentally, it is concluded that if the value of step is less than 300, then it will be heavy rain. Thus, if statement (27) is true, then statements (28) and (29) will display the message "HEAVY RAIN" in the LCD module for 2 seconds.

If the value of step is greater than or equal to 300, then statement (30) is true, and statements (31) and (32) will be executed to display the message "MODERATE or NO RAIN" in LCD module for 2 seconds.

5.5 ARDUINO-BASED MOBILE PHONE-CONTROLLED LIGHT

This project aims to control the switching of the bulb by using a mobile phone.

Readers can refresh the concepts by re-visiting the detailed explanation of relay and Bluetooth module given in Sections 3.11 and 3.19 of Chapter 3, respectively, and program related to relay and Bluetooth module in Sections 4.12 and 4.20 of Chapter 4, respectively.

The components required for this project are as follows:

i. Arduino UNO Board
ii. Generic breadboard
iii. 5 V relay
iv. Bluetooth module (HC-05)
v. Connecting wires
vi. Arduino cable.

Description of the Circuit:

The interfacing of the Bluetooth module (HC-05), 5 V relay board, and 220 V AC operated bulb with Arduino UNO board is shown in Figure 5.13. The VCC, GND (ground), TXD, and RXD pins of Bluetooth module are connected to

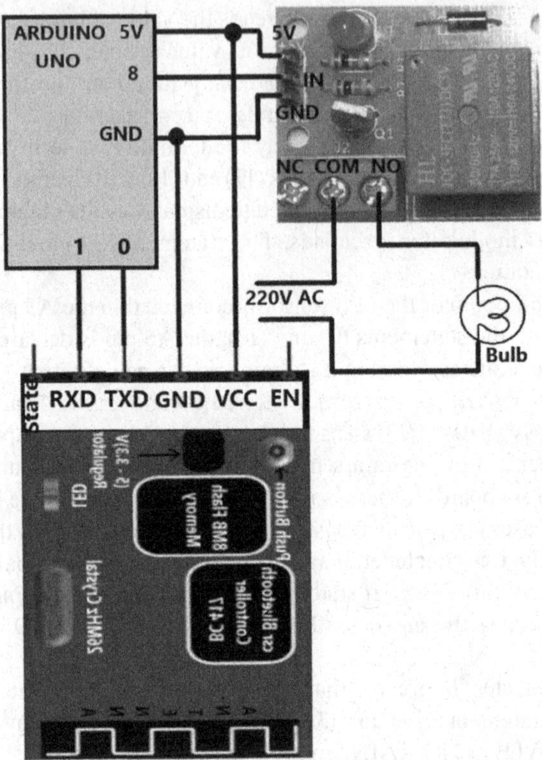

FIGURE 5.13 The circuit diagram to control the switching of the bulb by using a mobile phone.

5 V, GND (ground), RX (Pin 0), and TX (Pin 1) pins of Arduino board. The 220 V AC operated bulb is connected to NO (normally open) and COM (also called as pole) pins of relay. The 5 V and GND (ground) pins of relay are connected to 5 V and GND (ground) pins of Arduino board. The IN pin of relay is connected to pin number 8 of Arduino board. The Pin 8 of Arduino UNO is connected to the relay board to control the relay's triggering. The bulb is connected to the relay module, as shown in Figure 5.13. Under the default condition, since the bulb is connected between COM and NO terminals of relay, it will cause the bulb to turn off. If Pin 8 of Arduino generates 5 V, then the relay will be triggered, and this will cause the bulb to turn on.

The flowchart and the project program to control the bulb's switching by using a mobile phone are shown in Figures 5.14 and 5.15.

Description of the Program:

The statement (1) declares an integer-type variable "value". The Pin 8 of Arduino UNO is connected to the relay board to control the relay's triggering. Using the statements (2) and (3), the Pin 8 of Arduino is given a name "BULB" and

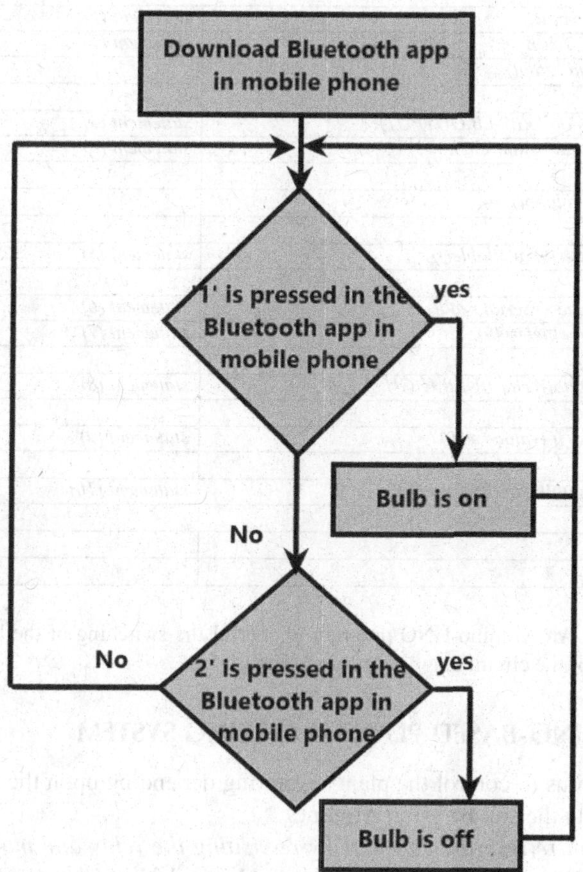

FIGURE 5.14 The flowchart of working principle of the project to control the switching of the bulb by using a mobile phone.

declared an output pin. The statement (4) initializes the serial communication between the Bluetooth module and Arduino board at 9,600 baud. If "1" or "2" is pressed in a mobile phone using Bluetooth app, then the statement (5) `if(Serial.available())` will return a true value and statement (6) will be executed and assigns either 49 or 50 to the "value" variable. If "1" is pressed in mobile phone, then "49" will be assigned to "value" variable, and if "2" is pressed in mobile, then "50" will be assigned to "value" variable.

If key "1" was pressed, then statement (7) will become true and statement (8) will be executed, causing 5 V output at Pin 8 of Arduino board. The 5 V at Pin 8 of Arduino board will trigger the relay and the bulb will turn on.

If key "2" was pressed, then statement (9) will become true and statement (10) will be executed, causing 0 V output at Pin 8 of the Arduino board. The 0 V at Pin 8 of the Arduino board will trigger the relay off, and the bulb will turn off.

int value;	*statement (1)*
int BULB=8;	*statement (2)*
void setup()	
{	
pinMode(BULB,OUTPUT);	*statement (3)*
Serial.begin(9600);	*statement (4)*
}	
void loop()	
{	
if(Serial.available())	*statement (5)*
{	
value = Serial.read();	*statement (6)*
if (value==49)	*statement (7)*
{	
digitalWrite(BULB,HIGH);	*statement (8)*
}	
else if (value==50)	*statement (9)*
{	
digitalWrite(BULB,LOW);	*statement (10)*
}	
}	
}	

FIGURE 5.15 An Arduino UNO program to control the switching of the bulb by using a mobile phone for the circuit diagram shown in Figure 5.13.

5.6 ARDUINO-BASED PLANT WATERING SYSTEM

This project aims to control the plant's watering depending upon the moisture content available in the soil by using Arduino.

Readers can refresh the concepts by re-visiting the relay and moisture sensing module's detailed explanation in Sections 3.11 and 3.21 of Chapter 3, respectively, and program related to relay in Section 4.12 of Chapter 4.

The components required for this project are as follows:

i. Arduino UNO Board
ii. Generic breadboard
iii. 5 V relay
iv. 5 V DC power supply
v. Soil moisture sensor (YL-69)
vi. Connecting wires
vii. Arduino cable
viii. 5 V DC water pump.

Description of the Circuit:

The interfacing of the soil moisture sensor (YL-69), 5 V relay board, and 20 V DC water pump with Arduino UNO board is shown in Figure 5.16. The 20 V DC water pump is connected to NO (normally open) and COM (also called as pole) pins of relay. The 5 V and GND (ground) pins of relay are

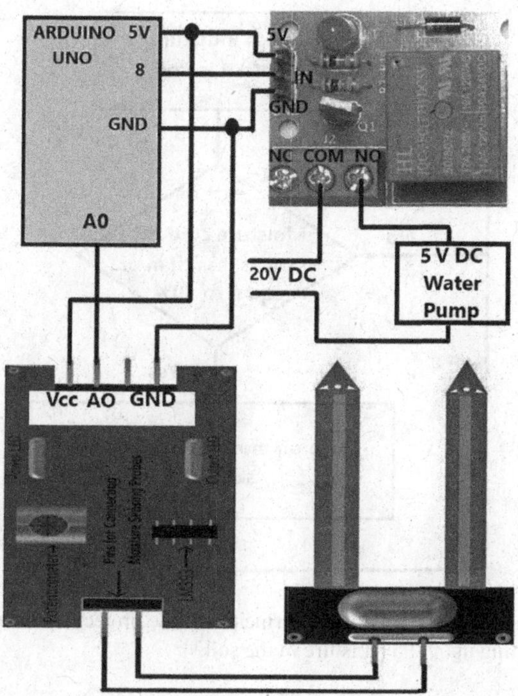

FIGURE 5.16 The circuit diagram to control the switching of water pump depending upon the moisture in the soil.

connected to 5V and GND (ground) pins of Arduino board. The IN pin of relay is connected to pin number 8 of Arduino board. The Vcc, GND (ground), and AO pins of soil moisture sensor are connected to 5 V, GND (ground), and A0 pins of Arduino board. The Pin 8 of Arduino UNO is connected to the relay board to control the triggering of relay. The 5V DC water pump is connected to the relay module, as shown in Figure 5.16. Under the default condition, since the water pump is connected between COM and NO terminals of relay, it will cause the water pump off. If Pin 8 of Arduino generates 5 V, then the relay will be triggered, and this will cause the water pump to on.

The moisture present in the soil is sensed by the soil moisture sensor, and depending upon the moisture content, the water pump will be turned on or off. The flowchart and the project program to control the water pump switching to water the plant by using Arduino are shown in Figures 5.17 and 5.18, respectively.

Description of the Program:

The statement (1) declares an integer-type variable "moisture_value". The Pin 8 of Arduino UNO is connected to the relay board to control the relay's triggering. Using the statements (2) and (4), Pin 8 of Arduino is given a name

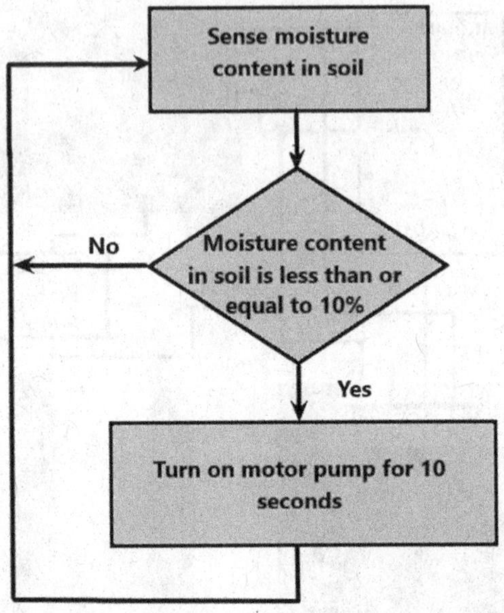

FIGURE 5.17 The flowchart of working principle of the project to control the switching of water pump depending upon the moisture in the soil.

int moisture_value;	*statement (1)*
int waterPump=8;	*statement (2)*
int analogInput=A0;	*statement (3)*
void setup()	
{	
pinMode(waterPump,OUTPUT);	*statement (4)*
pinMode(analogInput,INPUT);	*statement (5)*
}	
void loop()	
{	
moisture_value= analogRead(analogInput);	*statement (6)*
moisture_value =	*statement (7)*
map(moisture_value,550,10,0,100);	
if (moisture_value <=10)	*statement (8)*
{	
digitalWrite(waterPump,HIGH);	*statement (9)*
delay(5000);	*statement (10)*
}	
else	*statement (11)*
{	
digitalWrite(waterPump,LOW);	*statement (12)*
}	
}	

FIGURE 5.18 An Arduino UNO program to control turning on and off of water pump depending upon the moisture in the soil for the circuit diagram shown in Figure 5.16.

"waterPump" and declared an output pin. Using the statements (3) and (5), the A0 pin Arduino is given a name "analogInput" and declared an input pin.

The analog voltage proportional to the soil's moisture content will be generated from A0 pin of soil moisture sensor (YL-69) and connected to the analog input pin (A0) of the Arduino UNO board. Internally, A0 pin of Arduino is connected to a 10-bit analog-to-digital converter. The internal ADC of the Arduino UNO board has 1,024 steps ranging from 0 to 1,023. The statement (6) `moisture_value= analogRead(analogInput)` is used to assign 0–1,023 steps to integer "moisture_value" depending upon the analog voltage generated in response to the content of moisture in the soil. By experimentation, it is found that the soil moisture sensor (YL-69) value when the soil is dry and wet was 550 and 10, respectively. In order to map the soil moisture sensor (YL-69) value 550 and 10 in percentage from 0 to 100, the statement (7) `moisture_value = map(moisture_value, 550,10,0,100)` is used. Here, when soil is dry, 550 value of sensor will be mapped to 0%, and when soil is wet, 10 value of sensor will be mapped to 100%.

If the soil moisture content is less than or equal to 10%, the statement (8) will be true and statement (9) executes to trigger the relay, which will turn on the water pump. The water pump will remain on for 10 seconds due to statement (10).

If the soil moisture content is greater than 10%, the statement (11) will be true and statement (12) executes to turn off the water pump.

Appendix 1
Answers to Check Yourself

CHAPTER – 1

1. Integrated Development Environment
2. Sketch
3. setup() and loop()
4. 6
5. 14
6. 6
7. 13
8. 16 MHz
9. RISC-based architecture
10. 8 bit
11. 32 KB, 1 KB, and 2 KB
12. 28
13. LOW
14. A4 and A5 pins of Arduino UNO are used as SDA and SCL pins of I2C protocol.
15. SS/CS – Pin number 10 of Arduino UNO, MOSI – Pin number 11 of Arduino UNO, MISO – Pin number 12 of Arduino UNO, and SCK – Pin number 13 of Arduino UNO
16. Options b, c, and d are correct

CHAPTER – 2

1. SWITCH is the name assigned to Pin 13 of Arduino
2. Correct options are a and d
3. Correct option is d
4. 1 second or 1,000 ms
5. False
6. We cannot use 14 because we can write pin numbers from 0 to 13 only
7. In place of "Pinmode", correct keyword is "pinMode"

CHAPTER – 3

1. True
2. True
3. True
4. 10

5. Binary 0
6. All segments must be supplied with Binary 0
7. Logic 0
8. Logic 0
9. The alphabet “B” on mark “B10K” is the indication that the resistance of potentiometer varies linearly.
10. 2.5 V
11. 4.9 mV
12. True
13. 166
14. 3.25 V
15. True
16. True
17. 2
18. Motor does not rotate
19. Bulb will turn on
20. Bulb will turn off
21. False
22. Duty cycle = [on time/(on time + off time)] × 100%
    $= [150\times10^{-6}/(150\times10^{-6}+1.5\times10^{-3}\times1{,}000)]\times100$
    $= 9.0909\%$
23. Since the voltage drop across the LED is 1.4 V, from Ohm’s law (V = IR), we can write: $3.5–1.4 = 6\times10^{-3}\times R$. Thus, R = 350.
24. Correct options are a and b
25. Vout (in the absence of light) = 10 × 15 K/(15 K + 200 K) = 0.6976. Vout (in the presence of light) = 10 × 15 K/(15 K + 4 K) = 7.8947.

    Thus, option (a) is correct
26. For LM35, output voltage is linearly proportional to the temperature in degrees Celsius. The sensitivity of LM35 is 10 mV/°C. As the temperature increases, the output voltage also increases. The output voltage from LM35 can be directly connected with Arduino board (Analog pin), and we do not require any A/D converter.

    Thus, options (a) and (d) are correct.
27. When relay is not activated, the COM terminal is connected with NC, and when it is activated, the COM terminal is connected with NO output.

    Thus, options (a) and (d) are correct.
28. MQ2 gas sensor is most suitable for sensing gases like methane, smoke, etc. It generates analog output that is proportional to the intensity of the gas. It also has a digital port that can generate digital output, based on the threshold value set by a potentiometer.

    Thus, options (a) and (b) are correct.
29. False
30. False
31. False
32. False
33. True

References

1. Campbell, S. (2020, March 31). Basics of the SPI Communication Protocol Available: http:// www.circuitbasics.com.
2. Salab. (2013, June 5). Interface-DHT11-Using-Arduino Available: http:// www. www. instructables.com.
3. (2020). In-Depth: How MQ2 Gas/Smoke Sensor Works Available: http:// www. last-minuteengineers.com.
4. Arduino Reference Available: http:// www. www.arduino.cc.

Index

Note: page numbers in italics refer to figures and those in bold to tables.

Printed in the United States
by Baker & Taylor Publisher Services